RADIO RESOURCE MANAGEMENT IN CELLULAR SYSTEMS

THE KLUWER INTERNATIONAL SERIES
IN ENGINEERING AND COMPUTER SCIENCE

RADIO RESOURCE MANAGEMENT IN CELLULAR SYSTEMS

by

Nishith D. Tripathi
Nortel Networks

Jeffrey H. Reed
Virginia Polytechnic Institute and State University

Hugh F. VanLandingham
Virginia Polytechnic Institute and State University

KLUWER ACADEMIC PUBLISHERS
Boston / Dordrecht / London

ISBN 978-1-4419-4897-7   e-ISBN 978-0-306-47318-0

Distributors for North, Central and South America:
Kluwer Academic Publishers
101 Philip Drive
Assinippi Park
Norwell, Massachusetts 02061 USA
Telephone (781) 871-6600
Fax (781) 681-9045
E-Mail < kluwer@wkap.com >

Distributors for all other countries:
Kluwer Academic Publishers Group
Distribution Centre
Post Office Box 322
3300 AH Dordrecht, THE NETHERLANDS
Telephone 31 78 6392 392
Fax 31 78 6546 474
E-Mail < services@wkap.nl >

 Electronic Services < http://www.wkap.nl >

Library of Congress Cataloging-in-Publication Data

A C.I.P. Catalogue record for this book is available
from the Library of Congress.

Portions reprinted with permission, from [137], [138], [139], [140], and [141]
© 2010 IEEE and from [117] © 1997, TRLabs.

MATLAB® is a registered trademark of The MathWorks, Incorporated.

AT&T is a registered trademark of AT&T Corporation.

QUALCOMM is a registered trademark and registered service mark of QUALCOMM
Incorporated.

Printed on acid-free paper.

Printed in the United States of America.

To

Our Families

CONTENTS

CHAPTER 2   FUZZY LOGIC AND NEURAL NETWORKS

CHAPTER 3   ANALYSIS OF HANDOFF AND RADIO RESOURCE MANAGEMENT ALGORITHMS

CHAPTER 4   A GENERIC FUZZY LOGIC BASED HANDOFF ALGORITHM

CHAPTER 8   MICROCELLULAR HANDOFF ALGORITHMS

CHAPTER 9   OVERLAY HANDOFF ALGORITHMS

CHAPTER 10 SOFT HANDOFF ALGORITHMS

PREFACE

In a cellular communication system, a service area or a geographical region is divided into a number of cells, and each cell is served by an infrastructure element called the base station through a radio interface. Management of radio interface related resources is a critical design component in cellular communications. Efficient radio resource management cost-effectively enhances the capacity, i.e., the maximum number of users that can be supported in a given band, and quality of service perceived by users. Major radio resource management tasks are call admission control and resource control for ongoing calls. Call admission control involves control of both new calls and handoff calls. A new call is a call that originates within a cell and that requests access to the cellular system. A handoff call is a call that originated in one cell but requires and requests resources in another cell. Resource control for ongoing calls distributes the radio resources among existing users so that satisfactory quality of service, e.g., good voice quality and fast data retrieval from the Internet, is maintained. The radio resource management in a voice-centric cellular system is relatively simple. A voice call is admitted if there is any free channel, and speech quality is maintained by preserving a pre-determined signal to interference ratio through power control and handoff. However, emerging next generation cellular systems aim to serve both voice users and data users. The radio resource management in such systems is complex and must be designed carefully.

This book addresses the radio resource management in cellular systems from several perspectives. A comprehensive foundation of cellular systems, handoff, and basic radio resource management is provided. An introduction to two tools of artificial intelligence, neural networks and fuzzy logic, is given. These tools have been successfully used in signal processing and controls, and they can also be applied to develop efficient radio resource management algorithms. As a case study, novel handoff algorithms are designed using neural networks and fuzzy logic. These novel handoff approaches exploit attractive features of several existing algorithms, provide adaptation to the dynamic cellular environment, and allow systematic tradeoff among different system characteristics. To facilitate performance evaluation of handoff and overall radio resource management algorithms, simulation models are described. Note that the performance of a system carrying data is sensitive to data traffic characteristics. A set of data traffic models describing these characteristics is summarized. Since radio resource management is critical for next generation systems, major radio interface related features of three emerging cellular standards, IS-2000, UMTS, and 1xEV-DO, are discussed. Basic radio resource management considerations and approaches are described. This book provides a knowledge base of cellular systems (including emerging standards) and the application methodology of neural networks and fuzzy logic (analyzed for handoff) to enable the reader to develop high performance radio resource management algorithms for evolving cellular systems.

Chapter 1 investigates various aspects of handoff and overall radio resource management and includes an in-depth literature survey of handoff related research work. Desirable features and complexities of handoff are discussed. Several cellular system deployment scenarios that dictate certain handoff constraints are illustrated. Radio resource management in cellular systems is described in general terms.

Chapter 2 gives a brief introduction to the artificial intelligence related tools used in this research (artificial neural networks and fuzzy logic). A popular form of a fuzzy logic system is illustrated. Two neural networks, multi-layer perceptron and radial basis function network, are described.

Chapter 3 explains mechanisms used to evaluate handoff and radio resource management related performance of cellular systems. Several simulation models used in handoff research are described. A framework for radio resource management evaluation is provided along with data traffic models.

Chapter 4 proposes a new class of handoff algorithms that combines the attractive features of several existing algorithms and adapts the handoff parameters using fuzzy logic. Known sensitivities of handoff parameters are used to create a fuzzy logic rule base. The design procedure for a generic fuzzy logic based algorithm is outlined.

Chapter 5 suggests neural encoding of a fuzzy logic system to circumvent its large storage and computational requirements. The input-output mapping capability and compact data representation capability of neural networks are exploited to derive an adaptive handoff algorithm that retains the high performance of the original fuzzy logic based algorithm and that has an efficient architecture for storage and computational requirements.

Chapter 6 presents a fuzzy logic based algorithm with a unified handoff candidate selection criterion and adaptive direction-biasing. The unified handoff candidate selection criterion allows simultaneous consideration of several handoff criteria to select the best handoff candidate under given constraints. Enhanced direction-biasing is achieved by adapting the direction-biasing parameters.

Chapter 7 offers a new class of adaptive handoff algorithms that views the handoff problem as a pattern classification problem. Neural networks and fuzzy logic systems are good candidates for pattern classifiers due to their properties such as nonlinearity and to their generalization capability.

Chapter 8 explores handoff for microcells. Microcells impose distinct constraints on handoff algorithms due to the characteristics of the propagation environment. A generic adaptive algorithm suitable for a microcellular environment is proposed. Adaptation to traffic, interference, and mobility has been superimposed on the basic generic algorithm to develop another algorithm.

Chapter 9 analyzes handoff for overlay systems. An overlay system is a hierarchical architecture that uses large macrocells to overlay clusters of small microcells. Different handoff scenarios exist in an overlay environment, each with distinct objectives. This chapter proposes an adaptive overlay handoff algorithm that meets these objectives.

Chapter 10 focuses on soft handoff. Soft handoff exploits spatial diversity to increase signal energy for improved performance. A good soft handoff algorithm achieves a balance between the quality of the signal and the associated cost. This chapter highlights important considerations for soft handoff and develops adaptation mechanisms for soft handoff.

Chapter 11 discusses third generation and third-and-a-half generation cellular standards such as IS-2000, UMTS, and 1xEV-DO. Radio resource management aspects of these standards are highlighted.

We are grateful to our families for their love, warmth, and support. We thank our colleagues and reviewers who enhanced quality of the material in the book. We sincerely thank Magnus Almgren, Carey Becker, Ashvin Chheda, Ahmad Jalali, Farid Khafizov, Vinay Mahendra, David Paranchych, Bradley Stinson, Sirin Tekinay, Yiping Wang, Jim Weisert, Geng Wu, and Mehmet Yavuz for their support and comments. Technical expertise and efforts of these colleagues have improved the clarity and usefulness of the book, especially in the material related to emerging systems. Research environments at Virginia Tech and Nortel Networks provided a strong foundation for creation of this book. Lori Hughes deserves thanks for her prompt and efficient editing. We are thankful to Kluwer Academic Publishers for giving us an opportunity to publish our work. We hope that the readers will find the material in the book useful in their student endeavors or professional careers. The readers are encouraged to visit http://www.mprg.org for errata and other relevant information on the book.

Nishith D. Tripathi (Nortel Networks)
Jeffrey H. Reed (Virginia Tech)
Hugh F. VanLandingham (Virginia Tech)

Chapter 1

HANDOFF AND RADIO RESOURCE MANAGEMENT IN CELLULAR SYSTEMS

To a wireless user, dropped calls, blocked calls, and calls riddled with static are unacceptable. Users expect to remain connected while traveling through cells of a service area. Users also expect to be able to make calls at any time and to have good, clear connections. These user demands are addressed through the management of radio resources. Handoff is an important aspect of the radio resource management. Handoff makes a continuous connection possible by transferring a mobile user from one cell to another. Handoff also determines how many calls can be served in a given area and the quality perceived by users. Thus, efficient handoff algorithms are essential for preserving the capacity and quality of service of wireless communication systems.

1.1 INTRODUCTION TO HANDOFF

As a review, some of the terminology used in cellular communications follows [1].

- **Mobile Station (MS)**. The mobile station can be in motion at an unspecified location.
- **Base Station (BS)**. The base station is a fixed station used for radio communication with MSs.
- **Mobile Switching Center (MSC)**. The mobile switching center coordinates the routing of calls in a large service area. It is also referred to as the **Mobile Telephone Switching Office (MTSO)**.
- **Forward Channel**. The forward channel is the radio channel used for the transmission of information from the base station to the mobile station. It is also known as the **forward link** or **downlink**.
- **Reverse Channel**. The reverse channel is the radio channel used for the transmission of information from the mobile station to the base station. It is also known as the **reverse link** or **uplink**.

- **Handoff**. Handoff is a process of transferring support to a mobile station from a base station or from one channel to another. The channel change due to handoff occurs through a time slot for time division multiple access (TDMA), a frequency band for frequency division multiple access (FDMA), and a codeword for code division multiple access (CDMA) systems [2].
- **Cochannel Interference (CCI)**. The cochannel interference is caused when another signal in some remote cell is using the same channel as the desired signal.

The phases involved in the planning of cellular communications are assessment of traffic density, determination of cell sizes and capacity, selections regarding omni-directional or sectored cells and antenna directions, selection of best BS sites to cover the required area, allocation of frequencies, selection of power control parameters and handoff parameters [3].

This chapter carries out an in-depth investigation of the handoff aspects of cellular planning. The handoff process determines the spectral efficiency (i.e., the maximum number of calls that can be served in a given area [4]) and the quality perceived by users [3]. Efficient handoff algorithms cost-effectively preserve and enhance the capacity and Quality of Service (QoS) of communication systems [5].

Figure 1.1 shows a simple handoff scenario in which an MS travels from BS A to BS B. Initially, the MS is connected to BS A. The overlap between the two cells is the handoff region in which the mobile may be connected to either BS A or BS B. At a certain time during the travel, the mobile is handed off from BS A to BS B. When the MS is close to BS B, it remains connected to BS B.

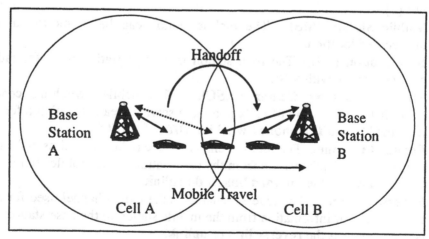

Figure 1.1: Handoff Scenario in Cellular Systems.

The overall handoff procedure can be thought of as having two distinct phases: the initiation phase (in which the decision about handoff is made) and the execution phase (in which either a new channel is assigned to the MS or the call is forced to terminate) [6]. Handoff algorithms normally carry out the first phase.

Handoff may be caused by factors related to *radio link, network management,* or *service options* [7, 8]. Network management and service related handoffs usually occur infrequently and easy to tackle. However, radio link related handoffs occur commonly and are most difficult to handle.

- **Radio Link Related Causes.** Radio link related causes reflect the quality perceived by users. Some of the major variables affecting the service quality are *received signal strength* (RSS), *signal-to-interference ratio* (SIR), and *system-related constraints*. Insufficient RSS and SIR reduce the service quality. Moreover, if certain system constraints (e.g., the synchronization requirement in a TDMA system) are not met, service quality is adversely affected.

- **Network Management Related Causes.** The network may hand off a call to avoid congestion in a cell. If the network identifies that the path used for information transfer is malfunctioning or is not the shortest one, it may hand off the call.

- **Service Options Related Causes.** When an MS asks for a service that is not provided at the current BS, the network may initiate a handoff so that the desired service can be offered [7]. A handover may also be initiated by the MS to connect to a service provider with a lower tariff.

A handoff made within the currently serving cell (e.g., by changing the frequency) is called an *intracell handoff*. A handoff made from one cell to another is referred to as an *intercell handoff*. Handoff may be hard or soft. *Hard handoff* (HHO) is "break before make," meaning that the connection to the old BS is broken before a connection to the candidate BS is made. HHO occurs when handoff is made between disjointed radio systems, different frequency assignments, or different air interface characteristics or technologies [9]. *Soft handoff* (SHO) is "make before break," meaning that the connection to the old BS is not broken until a connection to the new BS is made. In fact, more than one BS are normally connected simultaneously to the MS. For example, in Figure 1.1, both the BSs will be connected to the MS in the handoff region. Details of SHO are given in Section 1.9.3.

1.2   DESIRABLE FEATURES AND COMPLEXITIES OF HANDOFF

An efficient handoff algorithm can achieve many desirable features by trading different operating characteristics. Figure 1.2 summarizes the major desirable features of handoff algorithms, and several desirable features of handoff algorithms mentioned in the literature are described below [4, 5, 7, 10, 11, 12, 13, 14].

- Handoff should be fast so that the user does not experience service degradation or interruption. Service degradation may be due to a continuous reduction in signal strength or an increase in CCI. Fast handoff reduces CCI since it prevents the MS from going too far into the new cell.

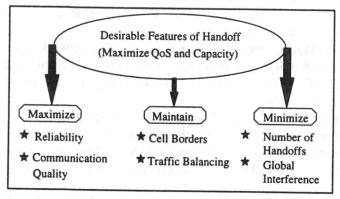

Figure 1.2: Desirable Features of Handoff Algorithms.

- Handoff should be reliable. This means that the call should have good quality after handoff. SIR and RSS help determine the potential service quality of the candidate BS.
- Handoff should be successful; a free channel should be available at the candidate BS. Efficient channel allocation algorithms and some traffic balancing can maximize the probability of a successful handoff.
- The effect of handoff on the QoS should be minimal. The QoS may be poor just before handoff due to a continuous reduction in RSS and SIR.
- Handoff should maintain the planned cellular borders to avoid congestion and high interference. Each BS can carry only its planned traffic load. Furthermore, there is a possibility of increased interference if the MS goes far into another cell site while still being connected to a distant BS since the distance between stations using the same channel is reduced and the distant BS tends to use a high transmit power to serve the distant MS.

- The number of handoffs should be minimized. Excessive handoffs lead to heavy handoff processing loads and poor communication quality. Minimizing the number of handoffs reduces the switching load. Unnecessary handoffs should be prevented; the current BS might be able to provide the desired service quality without interfering with other MSs and BSs.
- The target cell should be chosen correctly since there may be more than one candidate BS for handoff. Identification of a correct cell prevents unnecessary and frequent handoffs.
- The handoff procedure should minimize the number of continuing call drop-outs by providing a desired QoS.
- Handoff should have a minimal effect on new call blocking. For example, if many channels (called guard channels) are reserved exclusively for handoff, new call blocking probability will increase due to the reduction in the number of channels available for new calls.
- The handoff procedure should balance traffic in adjacent cells, eliminating the need for channel borrowing, simplifying cell planning and operation, and reducing the probability of new call blocking.
- The global interference level should be minimized by the handoff procedure. Transmission of bare minimum power and maintenance of planned cellular borders can help achieve this goal.

Existing handoff algorithms can give good performance only under certain situations due to complexities associated with handoff. Figure 1.3 shows the complexities associated with handoff.

- **Cellular Structure.** Different cellular structures or layouts put different constraints on handoff algorithms. Disjoint microcells and macrocells are expected to coexist in the cellular systems. In this case, microcells cover hot spots, while macrocells cover low traffic areas. Different radii cells require different handoff algorithm parameters to obtain good performance [15]. Some service areas may contain microcell-macrocell overlay in which microcells serve high traffic areas and macrocells serve high-speed users and overflow traffic. As the cell size decreases, the number of handoffs per call increases, the variables such as RSS, SIR, and bit error rate (BER) change faster, and the time available for processing the handoff requests decreases [16]. Moreover, the number of MSs to be handled by the infrastructure also increases.
- **Topographical Features.** A signal profile is characterized by the magnitude of the propagation path loss exponent and the breakpoint, i.e., the distance at which the magnitude of the propagation path loss exponent changes, and varies according to the terrain. The performance of a handoff algorithm depends on the signal profile in a region [17].

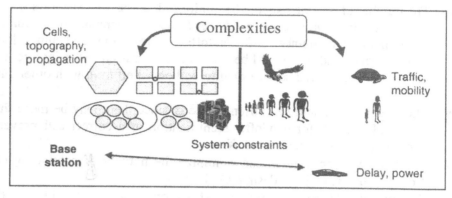

Figure 1.3: Complexities of Handoff.

- **Traffic.** In practice, traffic distribution is a function of time and space. Some of the approaches to cope with spatial nonuniformities of traffic are traffic balancing in adjacent cells, use of different cell sizes, nonuniform channel allocation in cells, and dynamic channel allocation [18].
- **Propagation Phenomena.** The radio propagation is strongly affected by surroundings. For example, due to a certain topological environment, the received signal strength can be higher at places distant from a BS than at places near the BS. Propagation characteristics in microcells are different from those in macrocells, e.g., the street corner effect [19]. In fact, it is shown in [13] that environment-dependent handoff parameters can give better performance than environment-independent handoff parameters.
- **System Constraints.** Some cellular systems are equipped with dynamic power control algorithms that allow the MS to transmit the least possible power while maintaining a certain quality of transmission. These systems coordinate power control and handoff algorithms to achieve their individual goals.
- **Mobility.** When an MS moves away from a BS at a high speed, the quality of communication degrades quickly. In such a case, handoff should be made quickly.

More importantly, *the evolution of a network is usually an on-going process*; new cells are gradually introduced, increasing the capacity to meet the demand. This network evolution necessitates adaptive resource management. The performance of growing cellular systems needs to be monitored and re-engineered frequently to maintain the QoS cost-effectively [20]. In summary, *to obtain high performance in the dynamic cellular environment, handoff algorithms should adapt to changing traffic intensities, topographical alterations, and the stochastic nature of the propagation conditions.*

1.3   CELLULAR SYSTEM DEPLOYMENT SCENARIOS

The radio propagation environment and related handoff and RRM challenges are different in different cellular structures.   Specific characteristics of the communication systems should be taken into account while designing handoff algorithms. Several basic cellular structures (such as macrocells, microcells, and overlay systems) and special architectures (such as underlays, multichannel bandwidth systems, and evolutionary architectures) are described next. Integrated cordless and cellular systems, integrated cellular systems, and integrated terrestrial and satellite systems are also described.

1.3.1 Macrocells

Macrocell radii are in several kilometers. The signal quality in the uplink and the downlink is approximately the same. The transition region between the BSs is large; handoff schemes should allow some delay to avoid flip-flopping. However, the delay should be short enough to preserve the signal quality because the interference would increase as the MS penetrates the new cell. This cell penetration is called *cell dragging*. Macrocells have relatively gentle path loss characteristics [5]. The averaging interval, i.e., the time period used to average the signal strength variations, should be long enough to get rid of fading fluctuations. First generation and second generation cellular systems provide wide area coverage using macrocells even in cities [19]. Typically, a BS transceiver in a macrocell transmits high output power with the antenna mounted several meters high on a tower to illuminate a large area. Figure 1.4 shows three clusters of seven cells in a macrocellular system.

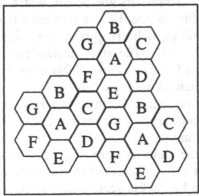

Figure 1.4: Seven-Cell Clusters in a Macrocellular System.

A cluster consists of a group of cells marked A through G. The available frequency band is divided into seven groups, with each group serving one cell. The frequency band is repeatedly used in each cluster. The use of seven cells in such geometry ensures that the co-channel distance is sufficiently long to maintain a desired signal to (co-channel) interference ratio for good signal quality. To achieve better SIR, the omni-directional cells shown in Figure 1.4 are often divided into three (or more) sectors. Such sectorization reduces the number of co-channel interferers and is widely used in deployed cellular systems.

1.3.2 Microcells

The use of microcells is an effective way of increasing the capacity of cellular systems.   Microcells increase capacity, but radio resource management becomes more difficult. Microcells can be classified as one-, two-, or three-dimensional, depending on whether they are along a road or a highway, covering an area such as a number of adjacent roads, or located in multilevel buildings, respectively [21]. Microcells can also be classified as hot spots (service areas with a higher traffic density or areas that are covered poorly), downtown clustered microcells (contiguous areas serving pedestrians and mobiles), and in-building 3-D cells (serving office buildings and pedestrians) [22]. The overlap region between the adjacent cells helps provide a seamless handoff. The required overlap puts a constraint on the smallest achievable cell size. Thus, there is a tradeoff between the size of the overlap region, or the quality of communication during handoff, and the capacity of the system [15].

Typically, a BS transceiver in a microcell transmits low output power with the antenna mounted at a lamp-post level, i.e., approximately 5 m above ground [19]. The MS also transmits low power, which leads to longer battery life and increased mobility. Since BS antennas have lower heights compared to the surrounding buildings, radio frequency (RF) signals propagate mostly along the streets [14, 23, 24]. The antenna may cover 100-200 m in each street direction, serving a few city blocks. This propagation environment has low time dispersion, which allows high data rates [15].

Microcells are more sensitive than macrocells to the traffic and interference due to short-term variations (e.g., traffic and interference variations), medium-/long-term alterations (e.g., new buildings), and incremental growth of the radio network (e.g., new base stations) [25]. The number of handoffs per cell is increased by an order of magnitude, and the time available to make a handoff is decreased [26]. Using an umbrella cell is one way to reduce the handoff rate. Due to the increase in the microcell boundary crossings and expected high traffic loads, a higher degree of decentralization of the handoff process becomes necessary [2]. The microcellular environment is highly interference-limited (i.e., noise is negligible and interference is a major concern) [2].

Microcells present the following constraints not typically found with macrocells [22]: (i) the amount of cabling must be reduced to enable the installation of several BS antennas at lamp-post level, (ii) a much denser cluster of wire lines and BSs is required, increasing the cost of infrastructure, (iii) the real estate is expensive in urban areas. These constraints pose stiff technical challenges to microcell engineering, and efficient resource management is required to achieve the maximum possible spectral efficiency.

Microcells encounter a propagation phenomenon called *corner effect*. The corner effect is characterized by a sudden large drop, e.g., 20-30 dB, in the signal strength, e.g., in 10-20 m distance, when a mobile turns around a corner. The corner effect is due to the loss of the line-of-sight (LOS) component from the serving BS to the MS. The corner effect demands a faster handoff and can change the signal quality very quickly. The rapid change in the RSS due to the corner effect affects the uplink more than the downlink in a microcellular environment. When a mobile turns a corner, the RSS at the MS becomes weaker. The uplink interference remains the same and downlink interference changes, potentially getting weaker [5]. A long measurement averaging interval is not desirable due to the corner effect. Moving obstacles can temporarily hinder the path between a BS and an MS, resembling corner effect.

In a microcellular system, there may be two types of handoff scenarios, an LOS handoff and a non-line-of-sight (NLOS) handoff. An LOS handoff is a handoff from one LOS BS to another LOS BS. An NLOS handoff is a handoff from an NLOS BS to an LOS BS. In an LOS handoff, premature handoff requests should be prevented. In an NLOS handoff, the handoff must be completed as quickly as possible as the user turns around the corner. Some of the solutions to deal with these different requirements for the LOS and NLOS handoffs in microcells are umbrella cells, macro-diversity, and switching to mobile-controlled handoff [5].

Gudmundson [27] studies the properties of symmetrical cell plans in a Manhattan-type environment. The symmetrical cell plans have four nearest co-channel BSs located at the same distance. Such cell plans can be classified into *half square* (Figure 1.5), *full square* (Figure 1.6), and *rectangular* (Figure 1.7) cell plans.

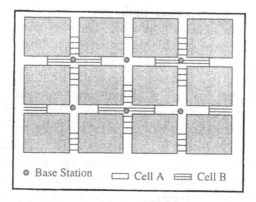

Figure 1.5: Half Square Cell Plan in a Microcellular System.

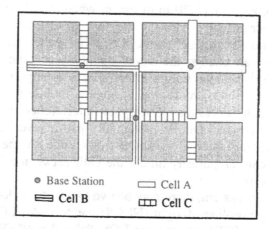

Figure 1.6: Full Square Cell Plan in a Microcellular System.

- **Half Square Cell Plan**. This cell plan places BSs with omni-directional antennas at each intersection, and each BS covers half a block in all four directions. This cell plan avoids the street corner effect and provides the highest capacity. This cell plan has only LOS handoffs.
- **Full Square Cell Plan**. A BS with an omni-directional antenna is located at every other intersection, and each BS covers a block in all four directions. It is possible for an MS to experience the street corner effect in this cell plan. The full square cell plan can have LOS or NLOS handoffs.
- **Rectangular Cell Plan**. Each BS, located in the middle of the cell, covers a fraction of either a horizontal or vertical street. This cell plan is scalable; fewer BSs can be used initially, and additional BSs can be added as the user density increases. The street corner effect is possible for this cell plan. This cell plan can have LOS or NLOS handoffs.

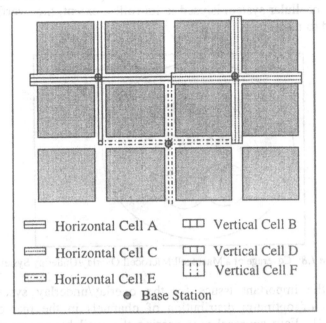

Figure 1.7: Rectangular Cell Plan in a Microcellular System.

1.3.3 Macrocell/Microcell Overlays

The congestion of certain microcells, the lack of service of microcells in some areas, and the high speed of some users are some reasons for higher handoff rates and signaling load for microcells [7]. To alleviate some of these problems, a mixed cell architecture (called an overlay/underlay system) consisting of large size macrocells (called umbrella cells or overlay cells) and small size microcells (called underlay cells) can be used. Figure 1.8 illustrates an overlay/underlay system. The macrocell/microcell overlay/underlay architecture provides a balance between maximizing the number of users per unit area and minimizing the network control load associated with handoff. Macrocells provide wide area coverage beyond microcell service areas and ensure better intercell handoff [17]. Microcells provide capacity due to greater frequency reuse and cover areas with high traffic density, i.e., *hot spots*. Examples of hot spots include an airport, a railway station, or a parking lot. In less congested areas, e.g., areas beyond a city center or areas outside the main streets of a city, traffic demand is not very high, and macrocells can provide adequate coverage in such areas. Macrocells also serve high speed MSs and the areas not covered by microcells, e.g., due to lack of channels or inadequate transmit power. Also, after the microcellular system is used to its fullest extent, the overflow traffic can be routed to macrocells.

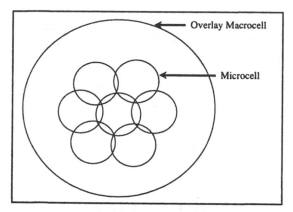

Figure 1.8: Coverage of a Macrocell/Microcell Overlay/Underlay System.

One of the important issues for the overlay/underlay system is the determination of optimum distribution of channels in the macrocells and microcells [28]. Four approaches to sharing the available spectrum between the two tiers are evaluated in [29]. Approach 1 uses TDMA for microcells and CDMA for macrocells. Approach 2 uses CDMA for microcells and TDMA for macrocells. Approach 3 uses TDMA in both tiers. Approach 4 uses orthogonal frequency channels in both tiers.

There are several classes of umbrella cells [30]. In one class, orthogonal channels are distributed between microcells and macrocells. In another class, microcells use channels that are temporarily unused by macrocells [31]. In yet another class, microcells reuse the channels already assigned to macrocells and use slightly higher transmit power levels to counteract the interference from the macrocells. Within the overlay/underlay system environment, four types of handovers need to be managed [32]: microcell to microcell, microcell to macrocell, macrocell to macrocell, and macrocell to microcell.

Combined *cell splitting* and *overlaying* is described in [33]. Reuse of channels in the two cells is achieved by establishing an overlaid small cell served by the same cell site as the large cell. Small cells reuse the split cell's channels due to the large distance between the split cell and the small inner cell while the large cell cannot reuse these channels. Combined cell splitting and overlaying is approximately 50% more spectrally efficient than *segmenting*, the process of distributing the channels among the small- and large-sized cells to avoid interference.

A practical approach for implementation of a microcell system overlaid with an existing macrocell system is proposed in [30], which introduces channel segregation, i.e., self-organized dynamic channel assignment, and automatic transmit power control to obviate the need to design channel assignment and transmit power control for the microcell system. The available channels are reused automatically between microcells and macrocells. A slight increase of transmit power for the microcell system compensates for the macrocell to microcell interference.

The methodology of the system based on the Global System for Mobile Communications (GSM) standard is extended to the macrocell/microcell overlay system in [34] that recommends the use of random frequency hopping and adaptive frequency planning, and different issues related to handoff and frequency planning for an overlay system are discussed.

Four strategies are designed to determine a suitable cell for a user for an overlay system in [35]. Two strategies are based on the dwell time (the time for which a call can be maintained in a cell without handoff), and the other two strategies are based on user speed estimation. A speed estimation technique based on dwell times is also proposed.

A CDMA cellular system can provide full connectivity between the microcells and the overlaying macrocells without capacity degradation. Shapira [22] analyzes several factors that determine the cell size, the SHO zone, and the capacity of the cell clusters. Several techniques for overlay/underlay cell clustering are also outlined.

Milstein [36] studies the feasibility of a CDMA overlay that can share the 1850-1990 MHz PCS band with existing microwave signals (transmitted by utility companies and state agencies). The results of several field tests demonstrate the application of such an overlay for the PCS band.

The issue of use of a CDMA microcell underlay for an existing analog macrocell is the focus of [37]. It is shown that high capacity can be achieved in a microcell at the expense of a slight degradation in macrocell performance. Grieco [37] finds that transmit and receive notch filters should be used at the microcell BSs. It shows that key parameters for such an overlay are the powers of the CDMA BS and MS transmitters relative to the macrocell BSs and the MSs served by the macrocells.

1.3.4 Special Architectures

There are several special cellular architectures that try to improve spectral efficiency without a large increase in the infrastructure costs. Some of these structures, discussed here, include an *underlay/overlay system,* which is different from an overlay/underlay system described earlier, and a *multichannel bandwidth system.* Many cellular systems are expected to evolve from a macrocellular system to an overlay/underlay system. A study that focuses on such evolution is also described here.

In an overlay/underlay system, frequency spectrum is divided between the macrocells and microcells in such a way that a macrocell uses certain channels throughout the cellular system. Also, the macrocell typically has a separate BS and a transmission tower. However, in an underlay/overlay system, a tighter reuse factor is used within an overlay. For example, assume that there are thirty six channels in a cluster of twelve cells. If there is no overlay or underlay, three channels will be available for each cell. In the conventional overlay/underlay system, two channels per cell can be used in a cluster of twelve microcells while the macrocell will use the remaining twelve channels throughout the cluster region. If uniform distribution of traffic is assumed, the effective number of channels per cell will still be three (two channels from a microcell and one channel from a macrocell). On the other hand, in one arrangement of an underlay/overlay scheme, two reuse factors, twelve and six, will be used instead of just one reuse factor, twelve, as shown in Figure 1.9.

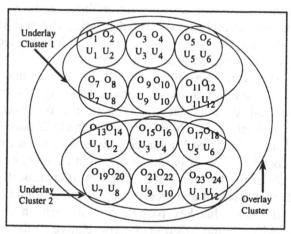

Figure 1.9: An Underlay/Overlay System.

Within a cluster of twelve cells, two channels per cell will be used in an overlay system, e.g., channels O_1 through O_{24} in Figure 1.9, and the remaining twelve channels will be distributed using the reuse factor of six, e.g., channels U_1 through U_{12} in Figure 1.9. Thus, within a single overlay cluster, there will be two underlay clusters, and each underlay cluster has a reuse factor of six. Hence, effectively there will be four channels per cell in an underlay/overlay system compared to three channels per cell for a non-underlay/overlay system. Further improvement in capacity can be obtained by using an even tighter reuse factor of three in an underlay cluster. In this case, there will be four underlay clusters within an overlay cluster. The overlay cluster uses two channels per cell, and the underlay cluster uses four channels per cell. Thus, effectively six channels per cell will be available. The underlay/overlay scheme can enhance capacity of the system without the infrastructure costs because the same BSs, transmission towers, and other hardware can be shared.

Multiple channel bandwidths can be used within a cell to improve spectral efficiency. In a multiple channel bandwidth system (MCBS), a cell has two or three ring-shaped regions with different bandwidth channels. Figure 1.10 shows an MCBS. Assume that 30 kHz is the normal bandwidth for a signal. Now, for a three-ring MCBS, 30 kHz channels can be used in the outermost ring, 15 kHz channels in the middle ring, and 7.5 kHz channels in the innermost ring. The areas of these rings can be determined based on the expected traffic conditions. Thus, instead of using 30 kHz channels throughout the cell, different bandwidth channels, e.g., 15 kHz, and 7.5 kHz, can be used to increase the number of channels in a cell. The MCBS uses the fact that a wide bandwidth channel requires a lower carrier to interference ratio (C/I) than a narrow channel bandwidth system for the same voice quality. For example, C/I requirements for 30kHz, 15 kHz, and 7.5 kHz channel bandwidths are 18 dB, 24 dB, and 30 dB, respectively, based on subjective voice quality tests [4]. If the transmit power at a cell site is the same for all the bandwidths, a wide channel can serve a large cell, while a narrow channel can serve a relatively small cell. Moreover, since a wide channel can tolerate a higher level of CCI, it can afford a smaller ratio of co-channel distance to cell radius (noted as D/R). Thus, in the MCBS, more channels become available due to multiple bandwidth signals, and frequency can be reused more closely in a given service region due to different C/I requirements.

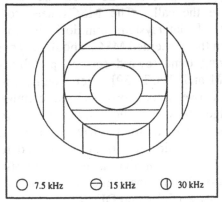

Figure 1.10: A Multiple Channel Bandwidth System.

Existing cellular systems are expected to evolve from large cells to small cells to cope with the increasing service demand. This type of evolution is the focus of [38], which considers three cell layout scenarios: the first layout has a non-layered cell architecture with macrocells; the second layout has a layered architecture with macrocells and medium-sized microcells; the third layout has macrocells and small microcells. For these layouts, the user penetration rate, i.e., the rate that the system can withstand to meet the QoS requirements, is estimated. A simulation model has also been developed to evaluate the performance of handoff algorithms. Different types of environments (such as domestic, office, streets) and different types of services (such as circuit switched voice, packet switched voice, and packet switched data services) have been taken into account.

1.3.5 Integrated Wireless Systems

Integrated wireless systems are exemplified by integrated cordless and cellular systems, integrated cellular systems, and integrated terrestrial and satellite systems. Such integrated systems combine the features of individual wireless systems to achieve the goals of improved mobility and low cost.

Terrestrial intersystem handoff may occur between two cellular systems or between a cellular system and a cordless telephone system. Examples of systems that need intersystem handoffs include GSM, Digital European Cordless Telephone (DECT), CDMA in macrocells, and TDMA in microcells. When a call initiated in a cellular system controlled by an MSC enters a system controlled by another MSC, intersystem handoff is required to continue the call [8]. In this case, one MSC makes a handoff request to another MSC to save the call. The MSCs need to have software for intersystem handoff if intersystem handoff is to be implemented. Compatibility between the concerned MSCs should be considered, too.

There is a growing trend toward service portability across dissimilar systems, such as GSM and DECT [39]. For example, it is nice to have an intersystem handoff between the cordless and cellular coverage. Cost-effective handoff algorithms for such scenarios represent a significant research area. This paper outlines different approaches to achieving intersystem handoff. Simulation results are presented for handoff between GSM and DECT. It is shown that a minor adjustment to the DECT specification can greatly simplify the implementation of an MS capable of an intersystem handoff between GSM and DECT.

In an integrated cellular/satellite system, advantages of satellites and cellular systems can be combined. Satellites can provide wide area coverage, completion of coverage, immediate service, and additional capacity (by handling overflow traffic). A cellular system can provide a high capacity economical system. Some of the issues involved in an integrated system are discussed in [40]. In particular, the procedures of the GSM are examined for their application to the integrated systems.

The future public land mobile telecommunication system (FPLMTS) will provide a personal telephone system that enables a person with a handheld terminal to reach anywhere in the world [41]. The FPLMTS will include low-earth-orbit (LEO) or geostationary-earth-orbit (GEO) satellites as well as terrestrial cellular systems. When an MS is inside the coverage area of a terrestrial cellular system, the BS will act as a relay station and provide a link between the MS and the satellite. When an MS is outside the terrestrial system coverage area, it will have a direct communication link with the satellite. Different issues such as system architecture, call handling, performance analysis of the access, and transmission protocols are discussed in [41]. The two handoff scenarios in an integrated system are examined here. While operating, the MS monitors the satellite link and evaluates the link performance. The RSSs are averaged, e.g., over a thirty second time period, to minimize signal strength variations. If the RSS falls below a certain threshold N consecutive times, e.g., $N = 3$, the MS begins measuring RSS from the terrestrial cellular system. If the terrestrial signals are strong enough, handoff is made to the terrestrial system, provided that the terrestrial system can serve the BS. Another scenario involves handoff from the terrestrial system to the Land Mobile Satellite System (LMSS). When an MS is getting service from the terrestrial system, the BS sends an acknowledge request at predefined intervals to ensure that the MS is still inside the coverage area. If an acknowledge request signal from the MS is not received at the BS for N consecutive times, it is handed off to LMSS.

Hu [42] focuses on personal communication systems with hierarchical overlays that incorporate terrestrial systems and satellite systems. The lowest level in the hierarchy is formed by microcells. Macrocells overlay microcells and form the middle level in the hierarchy. Satellite beams overlay macrocells and constitute the topmost hierarchy level. Two types of subscribers are considered, satellite-only subscribers and cellular/satellite dual subscribers. Call attempts from satellite-only subscribers are served by satellite systems, while call attempts from dual subscribers are first directed to the serving terrestrial systems with the satellites taking care of the overflow traffic.

1.4   HANDOFF CRITERIA

Several variables have been proposed and used as inputs, or *handoff criteria*, to handoff algorithms. The handoff criteria discussed here include RSS, SIR, distance, transmit power, traffic, call and handoff statistics, and velocity.

- **Received Signal Strength**. This criterion is simple, direct, and widely used. There is a close relation between the RSS and the distance between the BS and the MS. The lack of consideration of CCI is a disadvantage of this criterion. Several factors, e.g., topographical changes, shadowing due to buildings, and multipath fading, can cause the actual coverage area to be quite different from the intended coverage area. The RSS criterion can also lead to an excessive number of handoffs.

- **Signal-to-Interference Ratio**. An advantage of using SIR or C/I as a criterion is that SIR is a parameter common to voice quality, system capacity, and dropped call rate. BER is often used to estimate SIR. When actual C/I is lower than the designed C/I, voice quality becomes poor and the rate of dropped calls increases. SIR also determines the reuse distance. Unfortunately, C/I may oscillate due to propagation conditions and may cause the ping-pong effect (in which the MS repeatedly switches between the adjacent BSs). Another disadvantage is that even though BER is a good indicator of link quality, bad link quality may be experienced near the serving BS, and handoff may not be desirable in such situations [2]. In an interference-limited environment, deterioration in BER does not necessarily imply the need for an intercell handoff; an intracell handoff may be sufficient [17]. In [43], two methods for estimating raw channel BER over a Rayleigh fading channel are presented. Reference [41] proposes BER as a handoff criterion for an integrated system that consists of a cellular system and a terrestrial system. The BER is estimated from the faded signal.

- **Distance**. This criterion can help preserve the planned cell boundary. Distance can be estimated based on signal strength measurements [44] or delay between the signals received from different BSs [45]. Different research efforts have different opinions on the usefulness of this handoff criterion. Distance measurement can improve the handoff performance according to [11]; however, theoretical analysis in [46] does not consider the distance criterion better than others. If handoff occurs at the midway between two BSs, it distributes the channel utilization evenly [26]. Since future systems will require the location information of the MS, distance measurement will be available for use as a handoff criterion. The distance criterion may be useful for a macrocellular system, but it is prohibitive in a microcellular system since the precision of the distance measurement decreases with smaller cell sizes [2]. The determination of cell boundaries can avoid unnecessary handoffs. In the German cellular system C450, C/I and other data, such as signal strength, phase jitter, and the phase difference of the received digital signals, are measured and processed to detect the cell boundaries [47].

- **Transmit Power**. Transmit power as a handoff criterion can reduce the power requirement, reduce interference, and increase battery life.

- **Traffic**. Traffic level as a handoff criterion can balance traffic in adjacent cells. Rappaport [48] develops an analytical model for traffic performance analysis of a system and considers statistics of dwell times important.

- **Call and Handoff Statistics**. Statistics such as total time spent in the cell by a call and arrival time of a call in a cell can also be used as handoff criteria [7]. Elapsed time since last handoff is also a useful criterion since it can reduce the number of handoffs [8].

- **Velocity**. Velocity is an important handoff criterion, especially for overlay systems and velocity adaptive algorithms. Several algorithms use an estimate of velocity to modify handoff parameters. A method to adaptively change the averaging interval in a handoff algorithm for both small and large cells is presented in [49]. The method is based on the estimation of mobile velocity through maximum Doppler frequency f_D. This paper also outlines a method for estimating f_D from squared deviations of the signal envelope, both in Rayleigh fading and Rician fading environments. An adaptive scheme for optimal averaging has also been suggested.

1.5 CONVENTIONAL HANDOFF ALGORITHMS

Handoff algorithms are distinguished from one another in two ways, handoff criteria and processing of handoff criteria. Conventional handoff algorithms are described here, while emerging handoff algorithms are described in Section 1.6. Figure 1.11 summarizes the various types of handoff algorithms.

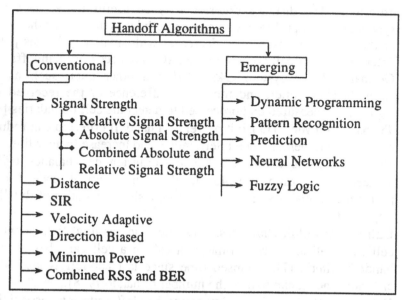

Figure 1.11: Handoff Algorithms at a Glance.

1.5.1 Signal Strength Based Algorithms

There are several variations of signal strength based algorithms, including *relative signal strength algorithms, absolute signal strength algorithms, and combined absolute* and *relative signal strength algorithms.* These algorithms are briefly discussed next.

According to the RSS criterion [14], the BS that receives the strongest signal from the MS is connected to the MS. The advantages of this algorithm are its ability to connect the MS with the strongest BS and the reduced processing load. However, the disadvantage is the excessive handoffs due to shadow fading variations associated with the signal strength. Another disadvantage is the MS's retained connection to the current BS, even if it passes the planned cell boundary, as long as the signal strength is above the threshold.

A variation of this basic RSS algorithm incorporates hysteresis. For such an algorithm, a handoff is made if the RSS from another BS exceeds the RSS from the current BS by an amount of hysteresis. The North American Personal Access Communication Systems (PACS) personal communication services (PCS) standard combines hysteresis with a dwell timer [5]. Hysteresis reduces the number of unnecessary handoffs but can increase drop-outs since it can also prevent necessary handoffs by introducing a delay in handoff [13]. A necessary balance between the number of handoffs and delay in handoff needs to be achieved by appropriate hysteresis and signal strength averaging. Corazza [6] describes a software simulator that allows the design of the number of samples to be averaged and the design of the hysteresis margin. The averaging should consider the MS speed and shadow fading. A scheme for estimating the shadow fading standard deviation based on squared deviations of the RSS at the MS is proposed in [50]. It is shown in [51] that the optimum handoff algorithm parameters are very sensitive to shadow fading standard deviation. To achieve robustness, more averaging and less hysteresis are required. However, to detect sudden changes in signal strength, e.g., due to street corner effect, less averaging and more hysteresis are required. To resolve this conflict, shadow fading deviation is estimated and used [49]. If the averaging interval is too short, fading fluctuates greatly. If it is too long, handoff is delayed. Thus, it is important to have an adaptive averaging interval. Dassanayake [52, 53] characterizes the variance of signal strength of the cell propagation environment and present its effect on handoff parameters such as signal averaging time and hysteresis. The simulation results based on GSM indicate that dynamic adjustment of propagation dependent handoff parameters could enhance the handoff performance. In a microcellular environment, a large hysteresis will avoid the ping-pong effect for LOS handoff while delaying NLOS handoff, which must be done very quickly to save the call [5]. The use of umbrella cells, microdiversity, and mobile-controlled handoff are some solutions to such contradictory goals [19]. The umbrella cell approach provides compatibility with existing systems.

For an absolute signal strength algorithm, when the RSS drops below a threshold level, handoff is requested. Typical threshold values are -100 dBm for a noise-limited system and -95 dBm for an interference-limited system [8]. Better handoff initiation can be obtained by varying the threshold. The threshold level should be varied according to the *path-loss slope L of the RSS* and the *level crossing rate (LCR) of the RSS*. If the slope L is high or *LCR* is high, the MS is quickly moving away from the BS, and, hence, handoff should be made fast (i.e., the handoff initiation threshold should be made higher in magnitude). If the slope L or *LCR* is low, the MS is moving slowly. So, handoff can be slow; the handoff initiation threshold can be made comparatively smaller. Thus, the mobile velocity and path-loss slope L can be used to determine the handoff initiation threshold dynamically such that the number of unnecessary handoffs is minimized and necessary handoffs are completed successfully. This algorithm has a serious disadvantage. When a threshold level is set based on the RSS, the following situations pose a problem [8]: (i) when RSS is high due to high interference, the handoff will not take place, although ideally, handoff is desirable to avoid interference; (ii) when RSS is low, handoff takes place even if voice quality is good, although ideally, such a handoff is not required. Some systems use supervisory audio tone (SAT) information with the RSS to avoid handoff.

A variation of the basic threshold algorithm is a two-level algorithm, which provides more opportunity for a successful handoff [8].

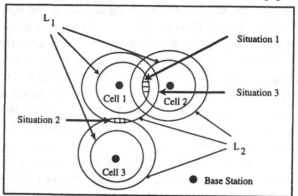

Figure 1.12: A Two-Level Handoff Algorithm.

Two handoff thresholds, L_1 and L_2, are defined with L_1 higher than L_2 as shown in Figure 1.12. When the RSS drops below L_1, a handoff request is initiated. If the MS is in a signal strength hole in the current cell or the candidate cell is busy, the possibility of handoff must be assessed. In this case, handoff is requested periodically, e.g., every five seconds. The handoff request is entertained only if the new RSS is stronger (Situation 1 in Figure 1.12). However, if the RSS reaches L_2, handoff will be made regardless of the relative RSS of the candidate BS (Situation 2 in Figure 1.12). Due to the two-level algorithm, the MS may come out of the hole or the candidate BS may have a free channel for handoff between L_1 and L_2. If a single threshold L_2 were used, the L_2 boundary might have been close to the candidate BS, causing interference (Situation 3 in Figure 1.12). However, in a two-level algorithm, L_1 boundaries of the BSs will have allowed handoff to be made earlier, avoiding high interference levels.

According to a combined absolute and relative signal strength algorithm, handoff takes place if the following two conditions are satisfied [54]: the average signal strength of the serving BS falls below an absolute threshold T dB, and the average signal strength of the candidate BS exceeds the average signal strength of the current BS by an amount of h (hysteresis) dB. The first condition prevents the occurrence of handoff when the current BS can provide sufficient signal quality. Beck [24] has shown that an optimum threshold achieves the narrowed handoff area (and hence reduced interference) and a low expected number of handoffs. Basic variables for this handoff algorithm are the length and shape of the averaging window, the threshold level, and the hysteresis margin [5]. Some of the findings of [15] for this algorithm are: (i) the probability of not finding a handoff candidate channel decreases as the overlap region increases; (ii) the probability of not finding a handoff candidate increases as the handoff threshold increases; (iii) the probability of a late handoff decreases as handoff threshold increases; (iv) the probability of unnecessary handoffs, i.e., the ping-pong effect, increases as handoff threshold increases; (v) the probability of unnecessary handoffs decreases as hysteresis increases.

1.5.2 Distance Based Algorithm

This algorithm connects the MS to the nearest BS. The relative distance measurement is obtained by comparing propagation delay times. This criterion allows handoff at the planned cell boundaries, giving better spectrum efficiency compared to the signal strength criterion [14]. German cellular system C450 uses this handoff criterion [55]. However, it is difficult to plan cell boundaries in a microcellular system due to complex propagation characteristics. Thus, the advantage of distance criterion over signal strength criterion begins to disappear for smaller cells due to inaccuracies in distance measurements.

1.5.3 SIR Based Algorithms

For toll-quality voice, SIR at the cell boundary should be relatively high, e.g., 18 dB for Advanced Mobile Phone System (AMPS) and 12 dB for GSM. However, a lower SIR may be used for capacity reasons since co-channel distance and cluster size, i.e., the number of cells per cluster, are small for lower SIR and channels can be reused more frequently in a given geographical region [8]. SIR is a measure of communication quality. This algorithm makes a handoff when the current BS's SIR drops below a threshold and another BS can provide sufficient SIR. Hysteresis can be incorporated in the algorithm. The lower SIR may be due to high interference or low carrier power. In either case, handoff is desirable when SIR is low. However, SIR-based handoff algorithms prevent handoffs near nominal cell boundaries and cause cell-dragging and high transmit power requirements [56]. In analog systems, measuring SIR during a call is difficult. Hence, sometimes interference power is measured before a call is connected, and combined signal and interference power is measured during the call.

Chuah [57] has suggested an uplink SIR-based algorithm for a power controlled system. Each user tries to achieve a target SIR. Handoff is made when the user's SIR drops below a threshold, which is normally less than the target SIR.

1.5.4 Velocity Adaptive Algorithms

Handoff requests from fast moving vehicles must be processed quickly. A handoff algorithm with short temporal averaging windows can be used to tackle fast users. However, the concept of a "short" averaging window is relative to the mobile speed. Thus, optimal handoff performance will be obtained only at one speed if the length of the averaging window is kept constant. A velocity adaptive handoff algorithm provides good performance for MSs with different velocities by adjusting the effective length of the averaging window [58]. A velocity adaptive handoff algorithm can serve as an alternative to the umbrella cell approach to tackle high-speed users if low network delay can be achieved, which can lead to savings in the cost of the infrastructure.

One of the velocity estimation techniques uses LCR of the RSS in which the threshold level should be set as the average value of the Rayleigh distribution of the RSS [59], requiring special equipment to detect the propagation dependent average receiver power. Kawabata [59] proposes a method of velocity estimation in a Rayleigh fading channel based on the velocity's proportionality to the Doppler frequency. The velocity estimation technique exploits diversity reception. If the MS is already using selection diversity, special equipment is not required for this method.

Velocity adaptive handoff algorithms for microcellular systems are characterized in [58]. Three methods for velocity estimation are analyzed: the level crossing rate method, zero-crossing rate method, and covariance approximation method. It is found that the spatial averaging distance required to sufficiently reduce the effects of fading depends on the environment. For sample averaging, sample spacing should be less than 0.5λ (half the wavelength). Usually, a spatial averaging distance of 20λ to 40λ is sufficient for microcells. A velocity adaptive handoff algorithm can adapt the temporal averaging window, i.e., a window with a certain time length, by either keeping the sampling period constant and adjusting the number of samples per window or vice versa.

1.5.5 Direction-Biased Algorithms

In an NLOS handoff, the MS experiences the corner effect as explained in Section 1.3.2. Hence, if the MS moves fast and is not handed off quickly enough to another BS, the call may be dropped. Connecting the fast moving vehicles to an umbrella cell is one solution, and using better handoff algorithms is another. A direction-biased handoff algorithm represents such an alternative solution [60]. Direction-biasing improves *cell membership properties* and handoff performance in LOS and NLOS scenarios in a multi-cell environment. A handoff algorithm is said to possess good cell membership properties if the probability that the MS is assigned to the closest BS is close to one throughout the call duration [60]. Improvement in cell membership properties leads to fewer handoffs and reduced interference.

The basic idea behind this algorithm is that handoffs to the BSs toward which the MS is moving are encouraged, while handoffs to the BSs from which the MS is receding are discouraged. This algorithm reduces the probability of dropped calls for hard handoffs, e.g., for TDMA systems. The algorithm also reduces the time a user needs to be connected to more than one BS for soft handoffs, e.g., for CDMA systems, allowing more potential users per cell.

A variation of the basic direction-biased algorithm is the preselection direction-biased algorithm [60]. If the best BS is a receding one and has a quality only slightly better than the second best BS that is being approached, the handoff should be made to the second best BS because it is more likely to improve its chances of being selected. This provides a fast handoff algorithm with good cell membership properties without the undesirable effects associated with large hysteresis.

1.5.6 Minimum Power Algorithms

A minimum power handoff algorithm that minimizes the uplink transmit power by searching for a suitable combination of a BS and a channel is suggested in [56]. This algorithm reduces call dropping but increases the number of unnecessary handoffs. To avoid a high number of handoffs, the use of a timer is suggested. First, the channel that gives minimum interference at each BS is found. Then, the BS that has a minimum power channel is determined. Mende [11] uses a power budget criterion to ensure that the MS is always assigned to the cell with the lowest path loss, even if the thresholds for signal strength or signal quality have not been reached. This criterion results in the lowest transmit power and a reduced probability of CCI.

1.5.7   RSS and BER Based Algorithms

An algorithm based on both RSS and BER is described in [26]. For RSS, a threshold is used for the current BS, and a hysteresis window is used for the target BS. For BER, a separate threshold is defined. The target BS can be either included or excluded from the handoff decision process. The latter scheme is used in GSM in which the mobile does not know the signal quality of the target BS. In principle, it is possible to measure BER of the control channel of the target BS. Three parameters considered in the simulations are RSS threshold, BER threshold, and RSS hysteresis window size [26]. In general, a low threshold value reduces the handoff request probability. The best threshold value is the average signal level at the mid-point between two BSs. However, due to the propagation environment, this threshold must be estimated for each base site. An RSS hysteresis delays handoff significantly. The higher the BER threshold, the earlier the handoff request. Moreover, if the BER threshold of the target BS is used, the handoff request is delayed. The handoff request probability differs significantly with location (or BS sites), showing that propagation characteristics are highly dependent on local terrain features and environment. From experimental results, it was found that the signal level and BER profiles varied significantly. RSS gives a direct indication of the received energy at the MS, while BER gives an indication of CCI and transmission quality. The effect of the RSS threshold level on handoff is opposite to that of BER since the gradient of the BER is opposite to that of RSS. If the gradient of the signal level is steep, the handoff region is less sensitive to a small variation in the threshold. Hysteresis is useful in preventing premature handoff requests if signal profiles are fluctuating. This is very useful in small site cells. In large site cells, hysteresis should be relatively small since it may introduce a delay in handoff initiation. Actual data, i.e., measured RSS and BER, is used in a software simulator that implements handoff schemes [26].

1.6   EMERGING HANDOFF ALGORITHMS

1.6.1   Dynamic Programming Based Handoff Algorithms

Dynamic programming allows a systematic approach to optimization. However, it is usually model dependent (particularly the propagation model) and requires the estimation of some parameters and handoff criteria, such as signal strengths. So far, dynamic programming has been applied to very simplified handoff scenarios only.

Handoff is viewed as a reward/cost optimization problem in [61]. RSS samples at the MS are modeled as stochastic processes. The reward is a function of several characteristics, e.g., signal strength, CIR, channel fading, shadowing, propagation loss, power control strategies, traffic distribution, cell loading profiles, and channel assignment. Handoffs are modeled as switching penalties that are based on resources needed for a successful handoff. Dynamic programming is used to derive properties of optimal policies for handoff. Simulation results show this algorithm to be better than a relative signal strength based algorithm.

Kelly [62] views signal strength based handoff as an optimization problem to obtain a tradeoff between the expected number of handoffs and the number of service failures, events that occur when the signal strength drops below a level required for an acceptable service to the user. An optimal solution is derived based on dynamic programming and is used for comparison with other solutions. The handoff problem is defined as a finite horizon dynamic programming problem, and an optimal solution is obtained through a set of recursive equations. This optimal solution is complex and requires *a priori* knowledge of the mobile trajectory. A locally optimal (or greedy) algorithm has been derived that uses the threshold level and gives a reasonable number of handoffs.

The handoff problem is formulated in a stochastic control framework in [63]. A Markov decision process formulation is used, and optimum handoff strategies are derived by dynamic programming. The optimization function includes a cost for switching and a reward for improving the quality of the call. The optimum decision is the hysteresis value representing the difference in RSSs from the BSs.

1.6.2 Pattern Recognition Based Handoff Algorithms

The handoff problem is formulated as a pattern recognition problem in [64]. Pattern recognition (PR) identifies meaningful regularities in noisy or complex environments. These techniques are based on the idea that the points that are close to each other in a mathematically defined feature space represent the same class of objects or variables. *Explicit PR techniques* use discriminant functions that define $n-1$ hypersurfaces in an n-dimensional feature space. The input pattern is classified according to their location on the hypersurfaces. *Implicit PR techniques* measure the distance of the input pattern to the predefined representative patterns in each class. The sensitivity of the distance measurement to different representative patterns can be adjusted using weights. Clustering algorithms and fuzzy classifiers are examples of implicit methods. The environment in the region near cell boundaries is unstable, and many unnecessary handoffs are likely to occur. The PR techniques can help reduce this uncertainty by efficiently processing the RSS measurements.

1.6.3 Prediction-Based Handoff Algorithms

Prediction-based handoff algorithms use the estimates of future values of handoff criteria, such as RSS. Kapoor [65] proposes this technique and shows it to be better than the relative signal strength algorithm and the combined absolute and relative signal strength algorithm via simulations. An adaptive prediction-based algorithm has been proposed to obtain a tradeoff between the number of handoffs and the overall signal quality [65]. Signal strength based handoff algorithms can use path loss and shadow fading to make a handoff decision. The path loss depends on distance and is determinate. The shadow fading variations are correlated and hence can be predicted. The correlation factor is a function of the distance between the two locations and the nature of the surrounding environment [66]. The proposed prediction based algorithm exploits this correlation property to avoid unnecessary handoffs. The future RSS is estimated based on previously measured RSSs using an adaptive FIR filter. The FIR filter coefficients are continuously updated by minimizing the prediction error. Depending upon the current value of the RSS (RSS_c) and the predicted future value of the RSS (RSS_p), the handoff decision is given a certain priority. Based on the combination of RSS_c and RSS_p, hysteresis may be added. The final handoff decision is made based on the calculated handoff priority.

1.6.4 Neural Handoff Algorithms

Most of the proposed neural techniques have shown only preliminary simulation results or have proposed methodologies without the simulation results. These techniques have used simplified simulation models. Learning capabilities of several paradigms of neural networks have not been utilized effectively in conjunction with handoff algorithms to date. Liodakis [2] presents a signal strength based handoff initiation algorithm using a binary hypothesis test implemented as a neural network. However, simulation results are not presented. A methodology based on an artificial neural network is proposed in [16]. Preliminary simulation results show that this methodology is suitable for multicriteria handoff algorithms.

1.6.5 Fuzzy Handoff Algorithms

A fuzzy handoff algorithm is proposed in [67]. The fuzzy handoff algorithm is shown to possess enhanced stability (less frequent handoffs). A hysteresis value used in a conventional handoff algorithm may not be enough for heavy fading, while fuzzy logic has inherent fuzziness that can model the overlap region between the adjacent cells, which is the motivation behind this fuzzy logic algorithm. A fuzzy classifier is used in [64] to process the signal strength measurements to select a BS to serve a call. The performance of this algorithm in a microcellular environment is evaluated. A handoff procedure using fuzzy logic is outlined in [68], which incorporates signal strength, distance, and traffic. Preliminary simulation results are presented. The concept of cell membership degree for handoff is explained in [12]. The methodology proposed in [12] allows systematic inclusion of different weight criteria and reduces the number of handoffs without excessive cell coverage overlapping. It is shown that the change of RSS threshold as a means of introducing a bias is an effective way to balance traffic while allowing few or no additional handoffs.

1.7 HANDOFF PRIORITIZATION

One way to reduce the handoff failure rate is to prioritize handoff. Handoff algorithms that try to minimize the number of handoffs give poor performance in heavy traffic situations [69]. In such situations, a significant handoff performance improvement can be obtained by prioritizing handoff.

1.7.1 Introduction to Handoff Priority

Channel assignment strategies with handoff prioritization have been proposed to reduce the probability of forced termination [70, 71]. Two basic methods of handoff prioritization, guard channels and queuing, are explained next.

- **Guard Channels**. Guard channels improve the probability of successful handoffs by reserving a fixed or dynamically adjustable number of channels exclusively for handoffs. For example, priority can be given to handoff by reserving N channels for handoffs among C channels in the cell [72]. The remaining C-N channels are shared by both new calls and handoff calls. A new call is blocked if the number of channels available is less than C-N. Handoff fails if no channel is available in the candidate cell. However, this concept has the risk of underutilization of spectrum. An adaptive number of guard channels can help reduce this problem. Efficient usage of guard channels requires the determination of optimum number of guard channels, knowledge of the traffic pattern of the area, and estimation of the channel occupancy time distributions.

- **Queuing of Handoff**. Queuing is a way of delaying handoff [8]; the MSC queues the handoff requests instead of denying access if the candidate BS is busy. Queuing new calls results in increased handoff blocking probability. The probability of a successful handoff can be improved by queuing handoff requests at the cost of increased call blocking probability and a decrease in the ratio of carried-to-admitted traffic since new calls are not assigned a channel until all the handoff requests in the queue are served. Queuing is possible due to the overlap region between the adjacent cells in which an MS can communicate with more than one BS. If handoff requests occur uniformly, queuing is not needed; queuing is effective only when handoff requests arrive in groups and when traffic is low. Conditional effectiveness of queuing has two causes: first, if traffic intensity is high, it is highly unlikely that a queued handoff request will be entertained, and secondly, when there is moderate traffic and when traffic arrives in bundles, a queued handoff request is likely to be entertained due to potential availability of resources in the near future and the lower probability of new handoff requests in the same period. Queuing is very beneficial in macrocells since the MS can wait for handoff before signal quality drops to an unacceptable level. However, the effectiveness of queuing decreases for microcells due to stricter time requirements. The combination of queuing and channel reservation can be employed to obtain better performance [73]. A long queuing delay is undesirable for real-time services such as voice.

1.7.2 Handoff Priority Schemes

Senarath [69] investigates the performance of different handoff priority schemes using a simulation model that incorporates transmission and traffic characteristics. The priority scheme of GSM has been evaluated. The simulation results show that the queuing and channel reservation schemes improve the dropout performance significantly, and the priority schemes provide up to 16% further improvement. Tekinay [71] presents a handoff prioritization scheme to improve the service quality by minimizing handoff failures and spectrum utilization degradation. If all the channels are occupied, new calls are blocked while handoff requests are queued. The handoff queue is dynamically reordered based on the measurements. The performance of the proposed handoff priority technique has been evaluated through simulations and compared with nonprioritized call handling and the first-in/first-out (FIFO) queuing scheme. The proposed scheme is shown to provide a lower probability of forced termination, a reduction in call blocking, a small reduction in traffic, and a small reduction in delay compared to the FIFO scheme under all traffic conditions. The newly proposed scheme improves the probability of forced termination at the cost of an increase in call blocking and a decrease in the ratio of combined-to-offered traffic. The priorities are defined by the RSS at the MS from the current BS. The degradation rate in service due to queuing depends on the velocity of the MS, and the proposed method considers this degradation rate. Gassvik [72] discusses two methods of giving priority to handoffs in a mobile system with *directed retry*, a feature of a cellular system, which allows the user to use a free channel in one of the neighboring cells [74]. Directed retry decreases the call blocking probability by sacrificing the handoff failure rate because there are fewer channels available for handoff in the candidate cell. Eklundh [74] presents simulation results of two handoff priority methods for a cellular system with directed retry.

1.8 HANDOFF AND RADIO RESOURCE MANAGEMENT

1.8.1 Introduction to Radio Resource Management

Examples of the RRM tasks performed by cellular systems include admission control, channel assignment, power control, and handoff [57, 75]. Chapter 11 discusses RRM for emerging cellular systems in detail. Major objectives of RRM are high capacity and throughput for a given quality of service. Interlinked radio resource management tasks are also possible. For example, handoff and channel assignment tasks can be combined [63]; a handoff request can be queued, and handoff is made when a channel becomes available. It should be noted that traditional cell planning may not be able to utilize the available spectrum efficiently due to highly environment-dependent radio propagation, rapid and unbalanced growth of radio traffic, and other factors [75]. The radio resource management tasks are explained next.

Admission Control. New calls and continuing calls can be treated differently. New calls may be queued. Handoffs may be prioritized. It is important to prevent the system from being overloaded. On the other hand, capacity is the revenue for service providers, and part of the perceived service quality can be attributed to the accessibility of the network. Queuing of calls is attractive especially for data services.

Channel Allocation. Tekinay [70] provides a tutorial on channel assignment (or allocation) strategies. Channel assignment strategies can be classified into fixed, dynamic, and flexible.

Fixed Channel Assignment (FCA) strategy permanently assigns a set of channels to each cell in a cluster. Some variations of the basic FCA strategy are the FCA with Borrowing (FCAB), the FCA with Hybrid Assignment (FCAHA), and the FCA with Borrowing-with-Channel-Ordering (FCABCO). In the FCAB, a channel can be borrowed from a neighboring cell if all the channels in a cell are busy (provided that this does not result in excessive interference). In the FCAHA, a set of channels in each cell is divided into two groups, one group is reserved for the local use and the other is kept for the lending purpose. FCABCO extends the idea of FCAHA by dynamically varying the ratio of the local-to-borrowable channels. Zhang [76] compares the performance of FCA and FCABCO with two proposed channel assignment strategies. Simulations for a 49-cell network have been carried out under uniform and nonuniform traffic conditions.

Dynamic Channel Assignment (DCA) strategy makes all the channels in a cluster available for use within a cluster. The actual channel assignment for a new call attempt is based on the minimization of a cost function that depends on future blocking probability, usage frequency of the candidate channel, and reuse distance of the channel. Dynamic channel allocation does not require *a priori* frequency planning but must determine whether co-channel usage is allowed or not. If adaptation to the changing propagation and interference conditions is done in a channel allocation algorithm, such an algorithm must guarantee a safe co-channel reuse distance. Hence, a measure of interference for the handoff candidate channel is required as an input to the channel allocation algorithm. Re [77] deals with dynamic channel allocation using a neural network. In microcells, the variations in the telephone traffic load are large compared to those in macrocells. Frodigh [78] proposes a DCA algorithm that adapts to these variations for a one-dimensional cellular system. The proposed algorithm maximizes the number of assigned calls and is suitable for distributed implementation. DCA gives better performance than FCA at low loads since it can adapt to traffic bursts. However, at high loads, DCA does not perform as well. Hence, some hybrid schemes have been suggested.

Flexible Channel Assignment (FLCA) strategy permanently distributes some channels among the cells in a cluster and keeps the remaining channels available for any cell's use when that cell's permanent channels are inadequate to cope with high traffic demand. As explained in Section 1.7.1, the use of guard channels exclusively for handoff requests results in under-utilization of the scarce channel resources. Narendran [79] presents a channel allocation algorithm that follows *most critical first policy* in which a free channel is assigned to the handoff request that would be the first to be cut-off if no channel were available at that time. Simulation results indicate that this algorithm is effective in reducing handoff failures. Goodman [80] describes signal strength based distributed channel assignment schemes for a one-dimensional cellular system.

Power Control. Power control is used to increase battery life, reduce health hazards, and contain interference. One way to exercise power control is to use SIR as a criterion. In this case, MSs try to attain a target SIR through continuous power adjustments. If the minimum possible power that meets the required C/I constraint at the receiver is transmitted, spectrum efficiency will increase compared to uncontrolled transmit power systems. Increasing transmit power to increase C/I for better transmission quality does not necessarily meet the objective since other transmitters in the system may also increase their power levels to reduce their interference, thus increasing the global interference level. Power control in both forward and reverse links is very critical for good performance of a CDMA system.

Handoff. One easy solution to BS assignment is to assign the MS to the nearest base station. However, due to the factors described in Section 1.2, the handoff issue becomes very complex. Intercell handoff can be viewed as an adaptive method of preserving the planned cell boundaries and subsequently reducing the interference. Adaptation to the spatial distributions of radio traffic (or interference) can be done by modifying cell areas and shapes dynamically by adapting the handoff parameters. This effect is called *cell breathing*.

1.8.2 Integrated Radio Resource Management Algorithms

Some algorithms that combine two or more radio resource management tasks are described next.

Channel allocation algorithms that adapt to the instantaneous interference and traffic situation can lead to an easier planning process. This is a tremendous advantage since the system grows stepwise with the traffic demand in most cases [25]. Beck [25] proposes a channel allocation algorithm that is adaptive to traffic and interference. It assumes that C/I of the candidate channels are measured periodically. C/I of the current channel is also measured for an existing call. The channel with the best C/I is selected for the new call. The candidate channel is accepted only if C/I of the candidate channel exceeds that of the old channel by some hysteresis value.

Chuah [57] proposes an integrated resource management based on four SIR thresholds. The resource management tasks incorporated into the algorithm are *admission control, power control, handoff*, and *channel allocation*. A call is dropped when SIR drops below γ_{drop}, e.g., 16 dB for AMPS. The threshold γ_{drop} is considered to be the minimum tolerable SIR for an acceptable speech quality. Power control is achieved by a target SIR threshold γ_t. Each MS tries to attain γ_t through power control. Call admission control is performed through a target SIR threshold γ_{new}. A new call attempt succeeds only if it can offer an SIR higher than γ_{new}. The SIR threshold γ_{new} ensures that the system is not packed too tightly. Otherwise, it may be difficult to find free channels for handoff. Furthermore, a new call, if admitted, will not cause severe interference to existing calls. Handoff and channel assignment are combined in the sense that handoff is made to the minimum interference channel when SIR drops below γ_{ho}.

The algorithm proposed in [75] uses RSS and transmission quality measure for the channels as handoff criteria. The BS allocation, channel assignment, and power control are treated in an integral manner. A new BS is selected in the case of new call setup and intercell handoff based on signal strength and possibly some network criteria. The comparison between the candidate BSs is done under equal transmit power levels. Power control is performed to increase spectral efficiency.

Yates [81] treats power control and BS assignment issues in an integral manner. The objective is to find a combination of BS assignment and transmit power to provide a feasible solution to the minimum transmit power (MTP) problem. An algorithm called minimum power assignment is proposed, which iteratively solves the MTP problem. During an iteration of the algorithm, the MS chooses a combination of BS and transmit power for which minimum power is needed to maintain an acceptable SIR (assuming that the other MSs transmit fixed powers at that time).

Hanly [82] also proposes a similar combined power control and BS selection algorithm to achieve higher capacity in a spread spectrum cellular system. The proposed algorithm adapts transmit powers of users and switches users between the BSs to minimize interference. The algorithm also reduces traffic congestion in a cell by moving the users to less congested adjacent cells.

1.9   HANDOFF PROTOCOLS

There are four basic types of handoff protocols: network controlled handoff (NCHO), mobile assisted handoff (MAHO), soft handoff, and mobile controlled handoff (MCHO). Figure 1.13 shows the tradeoff associated with the handoff protocols.   As the handoff decision making process is decentralized, i.e., moving from NCHO to MCHO, handoff delay (the time required to execute a handoff request) decreases, but the measurement information available to make a handoff decision also decreases.   These protocols are described next.

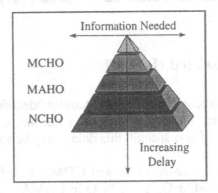

Figure 1.13: Handoff Delay and Measurement Information for Handoff Protocols.

1.9.1 Network Controlled Handoff

In the NCHO protocol, the network makes a handoff decision based on measurements of the RSSs of the MS at a number of BSs. The handoff command is sent on the voice channel by blanking the voice and sending data. Sometimes the network sets up a bridge connection between the old and new BS and thus minimizes the duration of handoff. In general, the handoff process (including data transmission, channel switching, and network switching) takes 100-200 ms and produces a noticeable click in the conversation. This click is imperceptible in a noisy voice channel; however, it is perceptible when handoff occurs at a reasonable signal quality [10]. Information about the signal quality for all users is located at a single point, i.e., the MSC. This information facilitates resource allocation. According to [83], the overall delay can be approximately five to ten seconds. This type of handoff is not suitable for a rapidly changing environment and a high density of users due to the associated delay.

This type of handoff is used in first generation analog systems such as AMPS [10]. Measurements are made only at BSs. RSSs of connected terminals are observed. SIR is measured by an SAT in AMPS. The BS transmits a tone with a frequency outside the audio range. This tone is echoed by the MS, and the BS estimates the interference from the quality of the received tone. Handoff is recognizable by the user as a clicking sound since a digital message is sent over an analog link. The measurements are made by two receivers at the BS [1]. The main receiver measures signal strengths of all its reverse voice channels. The locator receiver determines signal strengths of mobile users in neighboring cells. Based on the inputs from these receivers, the MSC decides whether handoff is necessary.

1.9.2 Mobile Assisted Handoff

The MAHO protocol distributes the handoff decision process. The MS makes measurements, and the MSC makes decisions. According to [83], there can be a delay of one second; this delay may be too much to counteract the corner effect.

The TDMA/FDMA based GSM and CDMA based Interim Standard-95 (IS-95) standards use MAHO. Since SHO in CDMA systems requires special processing, it is described separately in Section 1.9.3. Handoff related aspects of GSM are discussed here. An IS-95 based system uses SHO in conjunction with MAHO, and handoff related issues for such systems are discussed in Section 1.9.3. In GSM, Base Station Sub-system (BSS) includes a Base Station Transceiver Subsystem (BTS) and a Base Station Controller (BSC) [3]. The BTS is in contact with MSs through the radio interface and includes radio transmission and receiver devices as well as signal processing. The BSC is in contact with the network and is in charge of the radio interface management, mainly the allocation and release of radio channels and the handoff management. One BSC serves several BTSs, and several BSCs are connected to one MSC. The handoff time (the time between handoff decision and execution) in GSM is approximately one second [19].

Parameters used as handoff criteria in GSM are [3]

- static data such as maximum transmit power of the MS, the serving BTS, and the BTSs of the neighboring cells;
- real-time measurements performed by the MS such as the downlink transmission quality indicated by raw BER, downlink reception level on the current channel, and downlink reception levels from the neighboring cells;
- the BTS measurements (such as the uplink transmission quality quantified by raw BER, the uplink received level on the current channel, and the timing advance); and
- traffic considerations, cell capacity, and load.

GSM's TDMA structure consists of traffic channels (TCHs) and a number of control channels (CCHs). Since a fixed training sequence is transmitted in a certain time slot, the receiver can estimate its BER [11]. The BTS also measures the interference level on idle traffic channels. Measurement of the BTS-MS distance is an optional feature. The values measured by the MS are averaged and transmitted once in 480 ms over the SACCH (Slow Associated Control Channel) to the BTS. The BTS algorithm processes the 32 most recent measurements. The operators and manufacturers have complete freedom to implement their own handoff algorithms based on available parameters. According to [53], most manufacturers have designed their handoff algorithms based on signal strength. In this case, BER and the timing advance can act as alarm condition indicators rather than handoff algorithm inputs. Handoff to a different time slot on the same frequency channel is made for interference control reasons [84].

In GSM, handoff can be internal or external. If the serving and target BTSs are located within the same BSS, the BSC for the BSS can perform handoff without the involvement of the MSC. This is referred to as *intra-BSS handoff*. In an external handoff, the MSC coordinates the handoff. This type of handoff can further be classified as *intra-MSC* (within the same MSC) and *inter-MSC* (between MSCs) [9].

GSM based handoff algorithms are evaluated in [11, 32, 52, 53]. Dassanayake [53] is of particular interest and describes a GSM based handoff algorithm. The MS measures the RSS of the serving BS and the neighboring BSs through SACCH every 480 ms. These measurements are averaged over a certain period of time. The averaged signal strength is compared with a threshold. The MS also ranks the neighboring cells according to the magnitude of their RSSs. It compares the power budget criterion with the handover margin that helps avoid unnecessary handoffs. When the power budget criterion is met, a handoff is made to the top ranking neighboring cell if the RSS from that candidate cell is above the threshold. The threshold prevents handoff at low signal levels.

1.9.3 Soft Handoff

As mentioned earlier, the SHO is a special case of MAHO. The SHO is a "make before break" connection, i.e., the connection to the old BS is not broken until a connection to the new BS is made. SHO utilizes the technique of macroscopic diversity. Macroscopic diversity is a technique in which transmissions from an MS are received at different BSs and then used to obtain a good quality communication link [85]. The same concept can be used at the MS too. Macroscopic diversity is based on the principle of diversity combining that assumes that different BSs transmit and receive the same call with uncorrelated signal paths. Macroscopic diversity can, in some cases, provide good performance in terms of RSS and SIR for an interference limited system. Interference limited systems can exploit spatial diversity in the form of soft handoff. Macroscopic diversity is a form of spatial diversity and uses signals from several BSs to mitigate the effect of shadow fading (signal strength variations caused by buildings, foliage, and terrain features). It is shown in [86] that a four-branch macroscopic diversity can provide a 13 dB improvement in RSS and a 15 dB improvement in SIR (for a path loss exponent of four and a 10 dB shadow fading standard deviation).

Some of the diversity combining techniques include *selection diversity*, *maximal-ratio combining*, and *equal-gain combining* [85]. In selection diversity, the signal with the strongest SNR is selected. Maximum ratio combining uses co-phased signals within each receiver, and each signal is given a weighting factor according to its SNR before summation of the signals. Equal gain combining gives equal weight to all the signals before summation. The MSs must decode the signals from all base stations, which may be using the same or different channels. Such handoffs are called single channel SHO (SCSHO) and multiple channel SHO (MCSHO), respectively [83]. In the SCSHO, each participating base station transmits on the same channel. For high traffic situations, SCSHO may suffice. In case of the MCSHO, participating BSs use different channels. The mobile receiver decodes the signals on these orthogonal channels and achieves macroscopic diversity through diversity combining techniques. In general, signal quality will be better for MCSHO than SCSHO since the receiver has more degrees of freedom. However, at high traffic intensities, the MCSHO handoff scheme may not be feasible due to the unavailability of channels.

Ostling [83] investigates the effect of *simulcasting* in macrocellular and microcellular systems. Simulcasting is an example of SHO in which the MS is simultaneously connected to several BSs in the border region between the cells. The BSs participating in SHO simulcast, i.e., simultaneously transmit, replicas of the same signal to the MS.

There are several variations of SHO. The term *soft handoff* is used when old and new BSs belong to two different cells. The term *softer handoff* is used when the two signals correspond to the two different sectors of a sectorized cell [87]. When soft and softer handoffs occur simultaneously, the term *soft-softer handoff* is used. As far as the MS is concerned, there is no difference between SHO and softer handoff. For the network, additional hardware overhead is required for soft handoff. The handoff threshold needs to be small enough to bound the overall SHO percentage but large enough to allow efficient diversity combining. The MS needs more than one demodulator to exploit diversity combining techniques.

SHO can increase the capacity if exercised carefully. SHO increases the signal energy, enhancing the robustness to combat interference. This can lead to the reduction in the channel reuse distance, thereby increasing the capacity. However, SHO increases the level of co-channel interference, thereby reducing the system capacity. There is a tradeoff between these two conflicting factors. SHO has an advantage of changing SIR distribution. The MSs far from the base station receive more signal energy, and this reduces outage probability. Another advantage of SHO, increased signal energy reduces the switching of the call between the BSs. In particular, proper selection of the SHO region and its associated parameters can avoid the ping-pong effect common in hard handoff [85]. A disadvantage of SHO, the mobile undergoing SHO occupies channels of different BSs. Also, SHO tends to increase the traffic in the wired channels in a fixed network. The greater the number of BSs involved in SHO, the more traffic in the fixed network. Also, SHO requires an MS receiver capable of decoding multiple copies of the transmitted signal.

SHOs are common in CDMA. For IS-95, the BSs involved in SHO transmit on the same frequency, and the resultant signals are handled by the receiver as additional multipaths to be incorporated into the decoded signal [9]. The MS simultaneously communicates with multiple BSs. After the MS is firmly established in the new cell, the original BS disconnects the MS. Thus, CDMA based handoff provides a "make-before-break" switching function. Chapter 10 provides more details on SHO including SHO in IS-95 systems. Note that the CDMA systems forced to hand off to a different frequency use hard handoff. Hard handoff is a normal procedure for non-CDMA cellular systems. For example, FDMA and TDMA systems typically use hard handoff.

1.9.4   Mobile Controlled Handoff

In the MCHO, the MS is completely in control of the handoff process. This type of handoff has a short reaction time (on the order of 0.1 sec) and is suitable for microcellular system [83]. The MS does not have information about the signal quality of other users, and yet handoff must not cause interference to other users. The MS measures the signal strengths from surrounding base stations and interference levels on all channels. A handoff can be initiated if the signal strength of the serving base station is lower than that of another base station by a certain threshold. The MS requests the target BS for a channel with the lowest interference.

The MCHO is the highest degree of handoff decentralization. An advantage of decentralization of handoff, handoff decisions can be made quickly, and the MSC does not have to make handoff decisions for every mobile, which is a very difficult task for the MSC of high capacity microcellular systems [88].

The MCHO is used in the European standard for cordless telephones, DECT [19]. The MS and the BS monitor the current channel, and the BS reports measurements (RSS and BER) to the MS. The C/I of free channels are also measured. The handoff decisions are made by the MS. Both intracell and intercell handoffs are possible. The handoff time is approximately 100 ms.

1.10   MULTIPLE ACCESS SCHEMES IN CELLULAR SYSTEMS

Multiple access schemes allow multiple users to share the same system resources through orthogonal channels in a cell. Under ideal conditions, orthogonal channels eliminate in-cell interference. Examples of multiple access schemes are TDMA, FDMA, and CDMA. A TDMA system assigns different time slots to different users, while an FDMA system assigns different frequencies to users. In a CDMA system, different orthogonal spreading codes (e.g., Walsh codes) are allocated to users. To enable simultaneous communication between the BS and the MS, two approaches are widely used, frequency division duplexing (FDD) and time division duplexing (TDD). The FDD approach uses different frequency bands for the forward link and the reverse link, while the TDD approach uses the same frequency band but different time slots for the forward and reverse links. The GSM and IS-95 systems utilize FDD, and one mode of the emerging Universal Mobile Telecommunication System (UMTS) standard utilizes TDD.

In a cellular system, the available frequency spectrum is divided into several sets of channels and is repeatedly used in clusters of cells in a cellular system (see Section 1.3.1). Assume that a frequency spectrum with a 1.25 MHz bandwidth is available for the forward link of a system employing FDD. The 1.25 MHz bandwidth is divided into seven sets of channels for a 7-cell per cluster configuration and into four sets of channels for a 4-cell per cluster configuration. The reciprocal of the number of cells per cluster is called the *frequency reuse factor* [1]. The frequency reuse factor indicates how often the same frequency channel can be used in a given system. The larger the frequency reuse, the higher the system capacity. A minimum frequency reuse factor must be maintained to limit the CCI. An AMPS system typically has the frequency reuse factor of 1/7, while a GSM system typically has the frequency reuse factor of 1/7 or 1/4. A CDMA system has a typical frequency reuse factor of 1. The bandwidth occupied by a channel depends on the physical layer characteristics of the cellular system. For example, an AMPS system has a 30 kHz channel, and a GSM system has a 200 kHz channel. In a CDMA system, each user occupies the whole 1.25 MHz bandwidth, and the orthogonality between users is provided by Walsh codes. Thus, the CDMA system typically has a frequency reuse factor of 1; there is only one cell per cluster and each cell uses the same frequency.

The RF capacity of a TDMA/FDMA system on the forward link in a given cell can be easily determined based on the number of channels in the cell. The capacity of a CDMA system is discussed next. The number of Walsh codes is one factor that limits the forward link capacity of a CDMA system. In deployed CDMA systems, the capacity is found to be constrained by the forward link interference. Each user must be assigned sufficient power by the BS to overcome interference so that the target frame error rate (FER) is achieved. Most of the forward link interference in a cell served by a given BS is due to the power transmitted by other BSs and is thus a function of loading (the total power transmitted by the BS) in other cells. Since the forward link channels in a given cell or sector are orthogonal to each other, in-cell interference is zero (neglecting multipaths). Increasing the absolute amount of power at each BS will not increase capacity because each BS will eventually have to allocate more power to each user so that the user can overcome increased interference from other BSs that are now transmitting relatively higher powers. See Chapter 10 for more details on the capacity of a CDMA system. The CDMA technology arguably provides better RF capacity and is used in major third generation standards.

1.11 SUMMARY

A high performance handoff algorithm can achieve many desirable features by making appropriate tradeoffs. Handoff complexities and deployment scenarios were discussed. Different handoff criteria were analyzed. Both the conventional and emerging approaches for designing handoff algorithms were discussed. Handoff prioritization can improve handoff related system performance. Two basic handoff prioritization schemes, guard channels and queuing, were discussed. Handoff represents one of the radio resource management tasks carried out by cellular systems. Other resource management functions include admission control, channel assignment, and power control. Chapter 11 discusses radio resource management for emerging cellular systems. Handoff protocols were introduced. Basic multiple access schemes for cellular systems were summarized.

Chapter 2

FUZZY LOGIC AND NEURAL NETWORKS

There are several tools of artificial intelligence (AI) that help utilize human knowledge about the systems to enhance performance of the system. Some of the major AI tools are artificial neural networks (ANNs), fuzzy logic, genetic algorithms, and expert systems. This research exploits capabilities of neural networks and fuzzy logic to develop adaptive intelligent handoff algorithms. The focus of this chapter is on fuzzy logic systems and ANNs. Concepts of fuzzy logic theory and components of a popular fuzzy logic system (FLS) are discussed. Two ANN paradigms, multi-layer perceptron (MLP) and radial basis function network (RBFN), are introduced. The training process of these ANNs is explained.

2.1 INTRODUCTION TO FUZZY LOGIC

Information can be represented by numbers or linguistic descriptions. For example, *temperature* can be represented by the number 20°F or by the linguistic description "cold." The description "cold" is fuzzy and may represent any temperature between 10°F and 30°F, which can be called the *fuzzy set* (or *fuzzy region*) for the fuzzy variable temperature. Since humans usually think in terms of linguistic descriptions, giving these descriptions some mathematical form helps exploit human knowledge. Fuzzy logic utilizes human knowledge by giving the fuzzy or linguistic descriptions a definite structure.

A concise description of fuzzy logic theory is given next. First, basic concepts of fuzzy logic are introduced. These concepts are then utilized to explain a popular form of FLS that can serve as a building block in a system incorporating fuzzy logic. A comprehensive theory of fuzzy logic can be found in [89].

- **Fuzzy Set**. Let U be a collection of objects and be called *the universe of discourse*. A fuzzy set F U is characterized by a membership function $\mu_F(u):U \rightarrow [0,1]$ where μ_F represents the degree (or grade) of membership of $u \in U$ in the fuzzy set F. Figure 2.1 shows the membership functions of three fuzzy sets, "small," "medium," and "large," for a fuzzy variable *SIR*. The universe of discourse is all possible values of *SIRs*, i.e., $U = [15, 25]$. At an *SIR* of 19 dB, the fuzzy set "small" has the membership value 0.6. Hence, $\mu_{small}(19) = 0.6$. Similarly, $\mu_{medium}(19) = 0.4$, and $\mu_{large}(19) = 0$.

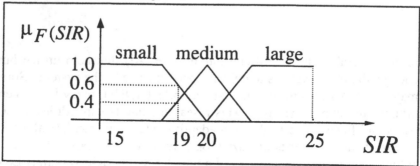

Figure 2.1: An Example of Fuzzy Logic Membership Function.

- **Support**. The support of a fuzzy set F is the crisp set of all points $u \in U$ such that $\mu_F(u) > 0$.

- **Center**. The center of a fuzzy set F is the point (or points) $u \in U$ at which $\mu_F(u)$ achieves its maximum value.

- **T-norm**. A T-norm, denoted by *, is a two-place function from [0,1] x [0,1] to [0,1], which includes fuzzy intersection, algebraic product, drastic product, and bounded product, defined as

$$x * y = min(x, y) \qquad \text{(fuzzy intersection)} \qquad (2.1)$$

$$x * y = xy \qquad \text{(algebraic product)} \qquad (2.2)$$

$$x * y = \begin{cases} x : y = 1 \\ y : x = 1 \\ 0 : x, y < 1 \end{cases} \qquad \text{(drastic product)} \qquad (2.3)$$

$$x * y = max(0, x + y - 1) \text{ (bounded product)} \qquad (2.4)$$

where $x, y \in [0,1]$.

- **Fuzzy Relation**. Let U and V be two universes of discourse. A fuzzy relation R is a fuzzy set in the product space $U \times V$, i.e., R has the membership function $\mu_R(u,v)$ where $u \in U$ and $v \in V$.

- **Sup-Star Composition**. Let R and S be fuzzy relations in $U \times V$ and $V \times W$, respectively. The sup-star composition of R and S is a fuzzy relation denoted by $R \circ S$ and is given by

$$\mu_{R \circ S} = sup_{v \in V} \left[\mu_R(u,v) * \mu_S(v,w) \right] \qquad (2.5)$$

where $u \in U$, $w \in W$, and * could be any operator in the T-norm defined earlier. It is clear that $R \circ S$ is a fuzzy set in $U \times W$.

- **Fuzzy Implications.** Let A and B be fuzzy sets in U and V, respectively. A fuzzy implication, denoted by $A \rightarrow B$, is a special kind of fuzzy relation in $U \times V$ with the following membership function:

$$\mu_{A \rightarrow B}(u,v) = \mu_A(u) * \mu_B(v). \qquad (2.6)$$

This fuzzy implication is known as fuzzy conjunction. Other types of fuzzy implications are also available [89].

Some of the popular FLS configurations include pure FLS, Takagi and Sugeno's fuzzy system, and Mamdani's fuzzy system [89]. The components of the FLS proposed by Mamdani are fuzzifier, fuzzy rule base, fuzzy inference engine, and defuzzifier as shown in Figure 2.2 [90].

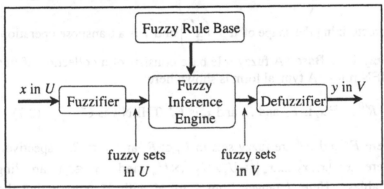

Figure 2.2: An Example of Fuzzy Logic System.

This configuration of the FLS (Mamdani FLS) has been widely used in industrial applications and consumer products. This FLS configuration has several advantages compared to other FLSs [89]. The Mamdani FLS has real-valued variables as its inputs and outputs, which is suitable for engineering applications where measured variables are real-valued and not fuzzy. Note that pure FLSs require fuzzy variables as inputs. The Mamdani FLS also provides a common framework, a rule base, for incorporating fuzzy IF-THEN rules to exploit human knowledge. Furthermore, this FLS allows several degrees of freedom in the selection of different components of the FLS. This FLS also allows fusion of numerical information and linguistic information. For example, numerical information, e.g., measurements, can be used to train the FLS to derive an adaptive FLS. The components of the Mamdani FLS are described next.

- **Fuzzifier**. The fuzzifier maps a crisp point, $\underline{x} = [x_1, x_2, ..., x_n]^T \in U$, into a fuzzy set $A' \in U$. Two choices for the fuzzifier forms are *singleton fuzzifier* and *nonsingleton fuzzifier*.

  Singleton Fuzzifier. A' is a fuzzy singleton with support $\underline{x}$, i.e., $\mu_{A'}\left(\underline{x}'\right) = 1$ for $\underline{x}' = \underline{x}$ and $\mu_{A'}\left(\underline{x}'\right) = 0$ for all other $\underline{x}' \in U$ with $\underline{x}' \neq \underline{x}$.

  Nonsingleton Fuzzifier. For this fuzzifier, $\mu_{A'}\left(\underline{x}'\right) = 1$, and $\mu_{A'}\left(\underline{x}'\right)$ decreases from 1 as $\underline{x}'$ moves away from $\underline{x}$. For example,

  $$\mu_{A'}\left(\underline{x}'\right) = exp\left[\frac{-\left(\underline{x}' - \underline{x}\right)^T \left(\underline{x}' - \underline{x}\right)}{\sigma^2}\right]$$ where σ is a parameter

  characterizing the shape of $\mu_{A'}\left(\underline{x}'\right)$ and T is a transpose operation.

- **Fuzzy Rule Base**. A fuzzy rule base consists of a collection of fuzzy IF-THEN rules. A typical form is shown here:

  $$R^{(l)}: \text{ IF } x_1 \text{ is } F_1^l \text{ and } ... \text{ and } x_n \text{ is } F_n^l, \text{ THEN } y \text{ is } G^l \qquad (2.7)$$

  where F_i^l and G^l are fuzzy sets in $U_i \subset R$ and $V \subset R$, respectively, and where $\underline{x} = [x_1, x_2, ..., x_n]^T \in U_1 \times U_2 ... \times U_n$ and $y \in V$ are linguistic variables. Here, l ranges from 1 to M with M representing the total number of rules. Let a fuzzy rule be expressed as "IF X is A, then Y is B" where X is the input fuzzy variable, Y is the output fuzzy variable, and A and B are corresponding fuzzy (linguistic) sets. A and B can be "small" and "big" respectively. The "IF" clause of the rule is called the *antecedent*, and the "THEN" clause the *consequent*. Each antecedent and consequent in a fuzzy logic rule forms a membership function that can be of different shapes, triangular and Gaussian shapes being more popular. Each input or output fuzzy variable has a membership degree of unity at the center value of the corresponding fuzzy set. Assume that the support of the fuzzy set A is between ten and thirty, and that twenty is the center value of the membership function for the fuzzy set A. Then, the input fuzzy variable X has a membership degree of unity with set A when X is twenty.

- **Fuzzy Inference Engine.** In the fuzzy inference engine, fuzzy logic principles are used to combine the fuzzy IF-THEN rules in the fuzzy rule base, and fuzzy sets in $U = U_1 \times U_2 \dots \times U_n$ are mapped into fuzzy sets in V. A fuzzy rule is interpreted as a fuzzy implication $F_1^l \times F_2^l \dots \times F_n^l \rightarrow G^l$ in $U \times V$. Let a fuzzy set $A' \in U$ be the input to the fuzzy inference engine. Then, each fuzzy IF-THEN rule determines a fuzzy set $B^l \in V$ using the sup-star composition:

$$\mu_{B^l}(u, v) = sup_{x \in U}\left(\mu_{F_1^l \times F_2^l \times \dots \times F_n^l \rightarrow G^l}(x, y) * \mu_{A'}(x)\right). \qquad (2.8)$$

Let $F_1^l \times F_2^l \dots \times F_n^l = A$ and $G' = B$. There are different interpretations for a fuzzy implication, and there are different T-norms as defined earlier. Hence, the above equation can be interpreted in a number of ways. One interpretation, called the *product-operation rule*, is shown here:

$$\mu_{A \rightarrow B}(u, v) = \mu_A(x) * \mu_B(y). \qquad (2.9)$$

This interpretation follows from the fuzzy conjunction implication by using the algebraic product for *.

Overall mapping of the fuzzy inference engine is described next. For an input A' (a fuzzy set in U), the output of the fuzzy inference engine can take two forms: (1) M fuzzy sets B^l ($l = 1, 2, \dots, M$) as in Eq. 2.8 with each one determined by one fuzzy IF-THEN rule as in Eq. 2.7, (2) one fuzzy set B', which is the union of the M fuzzy sets B^l. Thus,

$$\mu_{B'}(y) = \mu_{B^1}(y) \cup \dots \cup \mu_{B^M}(y). \qquad (2.10)$$

- **Defuzzifier.** The defuzzifier maps fuzzy sets in V into a crisp point, $y \in V$. One of the choices for the defuzzifier is center average defuzzifier, defined as

$$y = \frac{\sum_{l=1}^{M} \overline{y}^l \left(\mu_{B^l}\left(\overline{y}^l\right)\right)}{\sum_{l=1}^{M} \left(\mu_{B^l}\left(\overline{y}^l\right)\right)} \qquad (2.11)$$

where $\overline{y}^l$ is the center of the fuzzy set G^l, i.e., the point in V at which $\mu_{G^l}(y)$ achieves its maximum value, and $\mu_{B^l}\left(\overline{y}^l\right)$ is given by Eq. 2.5.

Application of the fuzzy logic concepts to a given problem (such as handoff) is explained in detail in Chapter 4.

2.2   INTRODUCTION TO NEURAL NETWORKS

Since the mid-1980s, researchers have been applying ANNs to solve a variety of problems including speech processing, pattern classification, multi-variable function interpolation, and signal prediction. ANNs are one tool of AI, others include fuzzy logic, genetic algorithms, and expert systems. An ANN is a massively parallel distributed processor that stores experimental knowledge; this knowledge is acquired by a learning process and is stored in the form of parameters of the ANN [91]. Characteristics of ANNs are massively parallel distributed architecture, ability to learn and generalize, fault tolerance, nonlinearity, and adaptivity. The learning in ANNs can be *unsupervised* or *supervised*. When an ANN undergoes learning in an unsupervised manner, it extracts the features from the input data based on a predetermined performance measure. When an ANN undergoes learning in a supervised manner, it is presented with the input patterns and the desired output patterns. The parameters of the ANN are adapted such that the application of an input pattern results in the desired pattern at the output of the ANN.

First, a fundamental component of the ANN, an artificial neuron, is explained. Using the model of the artificial neuron, two ANN architectures (or paradigms), multilayer perceptron and radial basis function network, are explained. The MLP and RBFN have been proven to be universal approximators [91], and hence they are used here to approximate the functional mapping between the FLS inputs and outputs. Training methods for the MLP and the RBFN to accomplish the task of function approximation are explained.

2.2.1   Fundamentals of ANNs

The ANN consists of a number of neurons arranged in a particular fashion. A nonlinear model of the artificial neuron is shown in Figure 2.3. The three basic elements of a neuron are the *synaptic weights* (or *weights*), the *summing junction*, and the *activation function*. Different activation functions include hard limit, linear, log-sig, and tan-sig. Threshold θ_k shown in Figure 2.3 can be considered as one of the weights with -1 as input. The weights and threshold (also known as *bias*) are referred to as parameters of the neuron.

The activation functions can be used with or without the bias. Usually, the ANN consists of more than one neuron.

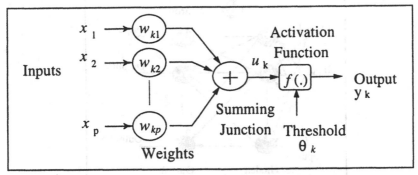

Figure 2.3: A Nonlinear Model of an Artificial Neuron.

The output of a neuron k is given by the following two equations:

$$u_k = \sum_{j=1}^{p} w_{kj} x_j \qquad (2.12)$$

$$y_k = f(u_k - \theta_k) \qquad (2.13)$$

where x_j ($j = 1, ..., p$) are the inputs, w_{kj} ($j = 1, ..., p$) are weights, θ_k is the threshold, $f(.)$ is the activation function, and y_k is the output of the neuron.

2.2.2 Paradigms of ANNs

Two paradigms of ANNs, MLP and RBFN, are briefly discussed next. Figure 2.4 shows a two-layer perceptron. Typically, the input layer is not counted as a separate layer since it does not do any processing. The first layer is called the hidden layer, and it consists of several neurons. The output layer consists of several neurons (equal to the number of outputs).

An MLP can be trained in a supervised manner by a very popular algorithm called the backpropagation algorithm. The training of an ANN requires a training data set that consists of input patterns and the desired output patterns. The backpropagation algorithm can be used to change the parameters of the perceptron to minimize the difference between the desired outputs (for given input patterns) and the actual outputs of the network. Details of the backpropagation algorithm can be found in [91]. The basic idea of the algorithm is explained next. Details of the training process of MLP and RBFN are given in Section 2.2.3.

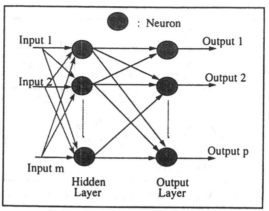

: Neuron

Input 1

Output 1

Input 2

Output 2

Output p

Input m

Hidden
Layer

Output
Layer

Figure 2.4: A Multilayer Perceptron.

The backpropagation algorithm consists of two distinct passes through the layers of the network, a forward pass and a backward pass. In the forward pass, an input pattern is applied to the input layer of the network, and it is processed by the parameters of the network. This produces a pattern at the output of the network. The output pattern is compared with the desired output pattern, and the error is calculated. In the backward pass, this error is propagated backward through the network, and the parameters of the network are modified by distributing this error among the parameters of the network. The forward and backward passes are made several times for all the training patterns. Gradually, the network begins to produce output patterns that resemble those desired. A basic backpropagation algorithm is very slow in convergence due to the requirements of small learning rates for stable learning. However, there are several techniques that can improve the speed and performance of the backpropagation algorithm, including Nguyen-Widrow weight initialization, use of momentum, and adaptive learning rate. It is shown in [92] that the weights generated with certain constraints result in a better function approximation. Use of Nguyen-Widrow initial conditions rather than random initial weights often reduces the training time by an order of magnitude. Momentum helps the network avoid getting stuck in shallow minima and leads to a better solution. Momentum can be included in the backpropagation algorithm by making the weight changes equal to the sum of a fraction (e.g., 0.95) of the last weight change and the new change suggested by the backpropagation learning rule. Thus, momentum allows a network to respond to both the local gradient and the recent trends in the error surface. Momentum allows the network to ignore small features in the error surface. An adaptive learning rate mechanism keeps the learning rate as high as possible while keeping the learning stable. The learning rate is adjusted based on the error performance. This research utilizes a two-layer perceptron with tan-sig activation functions in the hidden layer and linear activation functions in the output layer. The Neural Network Toolbox of MATLAB is used to train the two-layer perceptron.

The RBFN consists of three different layers, an input layer, a hidden layer, and an output layer as shown in Figure 2.5. The input layer acts as an entry point for the input vector; no processing takes place in the input layer.

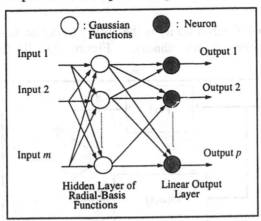

Figure 2.5: A Radial Basis Function Network.

The hidden layer consists of several Gaussian functions that constitute arbitrary basis functions (called radial-basis functions); these basis functions expand the input pattern onto the hidden layer space. This transformation from the input space to the hidden layer space is nonlinear due to nonlinear radial-basis functions. The output layer linearly combines the hidden layer responses to produce an output pattern. The rationale behind the working of the RBFN, a pattern-classification problem expressed in a high-dimensional space is more likely to be linearly separable than in a lower-dimensional space. The parameters of the RBFN are linear weights (in the output layer) and the positions and spreads of the Gaussian functions. A complete learning procedure can be found in [91]. Basically, in a supervised learning mode, these RBFN parameters are changed according to a gradient descent procedure that represents a generalization of the least-mean-squares (LMS) algorithm.

Two distinct phases of learning in the RBFN are the selection of centers of the radial basis functions and the determination of linear weights. Some of the methods for the selection of radial basis function centers are random selection (based on the training patterns), unsupervised selection, and supervised selection. Some of the methods for linear weight determination are pseudo-inverse memory and LMS algorithm. These weight determination methods find a mapping between the hidden unit space and the output layer. This research utilizes a three-layer RBFN. The Neural Network Toolbox of MATLAB is used to train the RBFN.

2.2.3 Training Methods for the MLP and the RBFN

At the very root of ANN utility is the ability to mimic unknown functions having some degree of smoothness. Figure 2.6 illustrates the basic configuration.

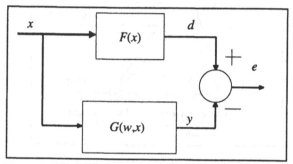

Figure 2.6: Basic Configuration for the Training of the ANN.

Figure 2.6 shows an unknown vector mapping F that generates an output d from a given input x. The block labeled with the function G represents an ANN structure with adjustable weights (parameters) w and input x. The design of the ANN consists of two parts: (1) choosing a particular structure, and (2) training the network with the available data.

Multi-Layer Perceptrons. By far, the most common type of ANN is the MLP, also known as a *feed-forward* network. The MLPs are configured in "layers" of processing elements (PEs), so that the inputs feed into one or more "hidden" layers of PEs that ultimately feed into the output layer. This type of network is classically trained using a gradient parameter update after each new data input/output pair has been applied. More details on the training algorithm will be given later in the section. Briefly, however, consider a certain structure, M-N-P, signifying M inputs (which would be known), N PEs in the hidden layer (N chosen by the designer), and P outputs (which also would be known), with the weights w initially set to random values. An input is applied to obtain the MLP output y and error e is formed as the difference between the "desired" output d and y. Knowing the MLP structure, a weight update equation can be derived for each parameter and takes the following form for each component (i^{th}) of w:

$$w_i(k+1) = w_i(k) - \eta \frac{\partial J}{\partial w_i} \qquad (2.14)$$

where $\eta \frac{\partial J}{\partial w_i}$ in Eq. 2.14 represents an increment to the ith weight that will be a step in the direction of the negative gradient so that the cost J will be reduced. The parameter η is the step size, also known as the learning rate. This parameter would typically be common to all weights at a given level.

Radial-Basis Function Networks. A second type of ANN considered here is the RBFN. As with any basis configuration, the output is a linear sum of basis functions. The most common basis function is the Gaussian function. Thus, an RBFN would process an input vector x as

$$y = \sum_{i=1}^{N} w_i \phi_i(x) \qquad (2.15)$$

where

$$\phi_i(x) = \exp\left(\frac{-(x-c_i)^2}{2\sigma^2}\right). \qquad (2.16)$$

Again, the weights w_i are to be adjusted from training data from the real mapping. However, since the weights occur linearly, they may be determined using matrix inverse methods. Note that the positions of the basis functions are distinct and, in some cases, even the spread parameter σ can be different for each basis function. The idea is to mimic the function at the different data points by using a (perhaps large) number of terms in Eq. 2.15 and Eq. 2.16.

MLP Training Methods. Consider the backpropagation method for training an MLP. Assuming for simplicity that the network has two inputs, N hidden layer nodes, and one output, the MLP function can be written as

$$y = b^{(2)} + \sum_{k=1}^{N} w_k^{(2)} y_k^{(1)} \qquad (2.17)$$

where

$$y_k^{(1)} = \Phi\left(b_k^{(1)} + w_{k1}^{(1)} x_1 + w_{k2}^{(1)} x_2\right). \qquad (2.18)$$

In this example, the nonlinear (squashing) functions $\Phi(x) = \tanh(x)$ are used only in the hidden layer. Note that the output layer parameters are labeled with a superscript (2) and those of the hidden layer by a superscript (1). The notation of Eq. 2.17 and Eq. 2.18 can be easily extended to more than two inputs. Now assume that the error is as shown in Figure 2.6 and that the intention is to minimize $J = e^2/2$. Using the chain rule of differentiation for the output layer bias parameter,

$$\frac{\partial J}{\partial b^{(2)}} = \frac{\partial J}{\partial e} \cdot \frac{\partial e}{\partial y} \cdot \frac{\partial y}{\partial b^{(2)}} = (e)(-1)(1) = -e. \qquad (2.19)$$

Similarly, for the other output layer parameters,

$$\frac{\partial J}{\partial w_k^{(2)}} = -e y_k^{(1)}. \qquad (2.20)$$

For the hidden layer parameters,

$$\frac{\partial J}{\partial b_k^{(1)}} = -e \, w_k^{(2)} \left(1 - y_k^{(1)}\right)\left(1 + y_k^{(1)}\right) \qquad (2.21)$$

$$\frac{\partial J}{\partial w_{kj}^{(1)}} = -e \, w_k^{(2)} \left(1 - y_k^{(1)}\right)\left(1 + y_k^{(1)}\right) x_j \text{ for } j = 1, 2. \qquad (2.22)$$

The backpropagation algorithm uses the above derivative information for a particular parameter, p, as follows: The update increment for p is given by

$$\Delta p^{(j)} = -\eta^{(j)} \frac{\partial J}{\partial p^{(j)}}. \qquad (2.23)$$

In summary, given the training data consisting of desired input/output pairs $\{x, d\}$ (see Figure 2.6), a training cycle can be described in terms of a forward and a backward pass.

Forward Pass. The input x is applied to the network (whose weights have been randomly initialized) resulting in the MLP output y. This output and all intermediate variables have thus been calculated, along with the error $e = y - d$ that will be used next.

Backward Pass. With the error and the various user-defined parameters, such as the learning rates, specified the derivatives in Eqs. 2.19 – 2.22 are now known. The adjustable weights can then be updated according to Eq.

2.23. This completes one "pattern" cycle of training. A useful variation of this training algorithm is to accumulate all updates over the training data before implementing any change in the parameters; this is called epoch training. A second variation is to include a momentum term. In this case, Eq. 2.23 is modified to include a component from the previous update as

$$\Delta p^{(j)}(i) = -\eta^{(j)} \left[\frac{\partial J}{\partial p^{(j)}} \right]_i + \mu^{(j)} \Delta p^{(j)}(i-1). \quad (2.24)$$

The new parameter, $\mu^{(j)}$, is the momentum rate for layer j. This variation helps to avoid getting stuck in a local minimum. The parameter i in Eq. 2.24 represents the update for pattern i.

Training Radial Basis Function Networks. Assume that the parameters of the ϕ functions have been chosen for the subsequent training of the w weights. Of course, the ϕ function parameters can be trained using a gradient method similar to that presented for the MLPs by obtaining the appropriate derivatives.

Batch Mode. Given the training data in the form of M input/output pairs, (x,d), for the ith input the desired output will be

$$d_i = w^T \phi(x_i). \quad (2.25)$$

Here, w and ϕ are vectors of the weights and basis functions, respectively. By combining all the data together,

$$D = w^T \Phi \quad (2.26)$$

where $D = [d_1, d_2, d_3, ..., d_M]$ and $\Phi = [\phi(x_1), \phi(x_2), \phi(x_3), ..., \phi(x_M)]$. With all the training compacted in this manner, the best fit for the vector of weights can be calculated by

$$w^T = D\Phi^T \left(\Phi\Phi^T \right)^{-1}. \quad (2.27)$$

The matrix Φ will have dimensions $N \times M$, where M is typically larger than N so that the matrix inverse in Eq. 2.27 will be for an $N \times N$ matrix.

Incremental Mode. There are situations in which the batch mode is inconvenient to implement, such as an on-line update. As an alternative, one may update the parameters in a manner similar to the back-propagation algorithm discussed above. Using the notation that $\Phi_j = \Phi(x_j)$, the RBFN output can be written as

$$y = \sum_{j=1}^{N} w_j \Phi_j . \qquad (2.28)$$

In a similar development to that of Eq. 2.19, the vector w can be updated following each pattern as

$$w^{new} = w^{old} + \eta \cdot e \cdot x . \qquad (2.29)$$

An improvement on the update of Eq. 2.29 is to normalize the input vector x by dividing by the square of its magnitude. In this case, it can be shown that the learning rate η should be chosen from the range 0 to 2, with η = 1 being the optimal choice when the data is not noisy.

Two ANNs, MLP and RBFN, are discussed from the point of view that the networks are to mimic an unknown vector mapping from the given data. In practice, the method may be extended to identify dynamic systems by including as inputs to the ANN various time delays of both the input and the output data. In this way, the ANN model becomes a type of nonlinear auto-regressive-moving-average model. Other applications may be in the area of pattern classification. The same training methods can be used with the modification that the desired output between two different classes would be designated by a (logic) 1 or 0. Chapter 7 shows how an ANN can be used as a pattern classifier to solve the handoff problem.

2.3   CONCLUSION

The tools of AI, such as neural networks and fuzzy logic, possess certain useful features such as nonlinearity, massive parallelism, learning capability, and human knowledge encoding capability. In particular, the research work reported in this book uses a full-fledged fuzzy logic system proposed by Mamdani. Two paradigms of neural networks, a multilayer perceptron and a radial basis function network, are utilized. Chapter 5 describes a procedure for utilizing an ANN in an application such as handoff.

Chapter 3

ANALYSIS OF HANDOFF AND RADIO RESOURCE MANAGEMENT ALGORITHMS

The performance analysis of handoff and radio resource management algorithms consists of two aspects, the performance metrics (or performance measures) and performance evaluation mechanisms. The performance metrics quantify the performance of algorithms. The evaluation mechanisms provide a means to collect statistics of performance metrics. Several important performance metrics are defined. Three major handoff evaluation mechanisms, analytical, simulation, and emulation, are briefly discussed. Basic constituents of simulation mechanisms for handoff are explained. Handoff simulation models used and developed for this research are described. Data traffic models that allow analysis of data systems are summarized. Simulation framework for the analysis of radio resource management is outlined.

3.1 SYSTEM PERFORMANCE MEASURES

Figure 3.1 depicts the handoff and RRM analysis procedure, which consists of two major components, performance evaluation mechanisms and performance metrics. These performance metrics are listed below.
- Call blocking probability is the probability that a new call attempt is blocked and is the ratio of the number of blocked calls and the number of call attempts.
- Handoff blocking probability is the probability that a handoff attempt is blocked and is the ratio of the number of blocked handoffs to the number of handoff attempts.
- Call dropping probability is the probability that an ongoing call is prematurely terminated.
- Handoff rate is the number of handoffs per unit of time.
- Handoff delay is the time interval between the initiation of a handoff request and the execution of the handoff request.

- Interference probability is the probability that the SIR is less than a threshold (often called the protection ratio) [14].
- Assignment probability is the probability that the MS is connected to a particular BS [14].
- Power limiting probability is the probability that the total required base station power exceeds the maximum power limit. Statistics of the degree of and the time duration of the power limiting are also useful.
- Packet delay is the delay experienced by a data packet. One definition of packet delay is the time difference between the packet arrival at the transmit buffer and the packet reception at the receive buffer. A 90-percentile packet delay indicates that 90% of the packets experience a delay less than a specified value.
- Throughput is the total number of bits transmitted in a sector (cell) per second (unit time).

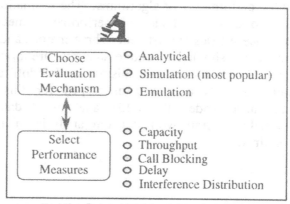

Figure 3.1: Procedure for the Analysis of Handoff and Radio Resource Management.

Important performance measures for handoff are handoff blocking probability, call dropping probability, handoff rate, interference probability, and assignment probability. Important performance measures for RRM are call blocking probability, handoff blocking probability, probability of power limiting, packet delay, and throughput. Some performance metrics may not be easily measurable in a given simulation model. In such cases, it is possible to infer a performance metric based on some other indirect measure. For example, call blocking probability and handoff blocking probability can be inferred from the cumulative distribution function (CDF) of traffic, i.e., number of calls in a cell, if a simulation model does not explicitly give these probabilities. The average number of handoffs made during a travel can give an idea about the handoff rate. CDFs of RSS and SIR give an indication of the call dropping probability.

3.2 HANDOFF EVALUATION MECHANISMS

Three basic mechanisms used to evaluate the performance of handoff algorithms include the analytical approach, the simulation approach, and the emulation approach.

3.2.1 Analytical Approach

This approach can quickly give a preliminary idea about the performance of some handoff algorithms for simplified handoff scenarios under specified constraints, e.g., assumptions about the RSS profiles.   Actual handoff procedures are quite complicated, and they are not memoryless. This makes the analytical approach less realistic. Real world situations make this approach complex and mathematically intractable. Some of the analytical approaches appearing in the literature are briefly touched upon here.

The level crossings of the difference between the RSSs from two BSs are modeled as Poisson processes for stationary signal strength measurements in [51]. This analytical work was extended to nonstationary signal strength measurements in [93], and the level crossings were modeled as Poisson processes with time-varying rate functions. The results in [51] and [93] are useful for determining the averaging interval and hysteresis level to achieve an optimum balance between the number of unnecessary handoffs and the delay in handoff for a simplified scenario in which an MS travels along a straight line from one BS to another at a constant velocity. Austin [94] incorporates the effect of CCI in the signal strength based handoff algorithm analysis presented in [51]. Zhang [54] develops an analytical model for analyzing performance of handoff algorithms based on both absolute and relative signal strength measurements and compares analytical results with simulation results.

Gudmundson [88] derives bounds for some performance measures and gives analytical expressions for the performance measures for a particular (linear) class of algorithms. Linear handoff algorithms do not use hysteresis and use only one quality measure, i.e., signal strength.

The effect of handoff techniques on cell coverage and reverse link capacity for a spread spectrum CDMA system is investigated in [95]. The paper shows that SHO increases both the cell coverage and reverse link capacity significantly compared to conventional hard handoff and derives quantitative performance improvement measures for cell coverage and capacity of the reverse link.

Prioritized handoff schemes are analyzed in [96]. It was assumed that the probability density function (pdf) of the speeds of cell-crossing terminals is the same as the pdf of the terminal speeds in cells. Reference [97] derives a more precise pdf using biased sampling in boundaries. The resultant analysis is computationally less complex and more accurate compared to the approach in [96].

An analytical model is proposed in [98] to study the traffic performance of a microcell/macrocell overlay for a PCS architecture. If a call cannot be served by a microcell, it is connected to a macrocell. The call is blocked if no channel is available in the macrocell. The overflow traffic to the overlay macrocell is computed. The residual time distribution for a macrocell is derived based on the assumed residual time distribution for a macrocell. The call termination probability for the macrocell is computed using the overflow traffic as input.

Teletraffic performance of a highway microcellular system with a macrocell overlay is presented in [99], assuming a TDMA scheme with ten channels per carrier and one carrier per BS. The teletraffic analysis assumes that the mobile speeds follow truncated Gaussian distribution. The probability of new call blocking and handoff call forced termination have been evaluated for three scenarios: when no priority is given to any MS, when priority is given to handoff calls, and when a macrocell overlay makes channels available to transfer calls from the MSs that would be blocked during a microcellular handoff.

The teletraffic analysis of a hierarchical cellular network (in which umbrella cells accept handoff requests that cannot be managed by microcells) is the focus of [100]. The handoff flow from a microcell to a macrocell is modeled as a Markov modulated Poisson process, and call blocking and call dropping probabilities are calculated.

3.2.2 Simulation Approach

The simulation approach is the most commonly used handoff evaluation mechanism. Several simulation models suitable for evaluation of different types of handoff algorithms under different deployment scenarios have been proposed and used in the literature. Usually, the analytical studies of handoff algorithms consider handoff between two BSs. However, the simulation approach allows incorporation of many features of a cellular system and a cellular environment into the evaluation framework. This approach provides a common testbed for comparison of different handoff algorithms. This approach also provides insight into the behavior of the system [5]. Field measurements are useful, but they are time-consuming and expensive. Software simulation provides fast, easy, and cost-effective evaluation. Simulation models usually consist of one or more of the following components: cell model, propagation model, traffic model, and mobility model. These components are described first, and specific simulation models are discussed next. Figure 3.2 shows the components of a typical simulation model.

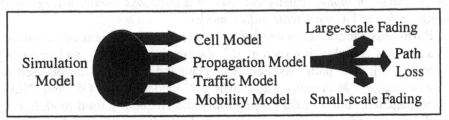

Figure 3.2: Simulation Model Components.

Cell Model. Cell planning strategies differ in microcells and macrocells. Cells can be considered as circles while considering handoff between two BSs in a neighborhood of two, three, or four cells. A macrocellular system is sometimes simulated as a 49-cell toroidal system that has seven-cell clusters with uniformly distributed traffic. Such a model can be used for a system with a frequency reuse factor of 1/7. Hexagonal cells are commonly used to model one or more tiers of cells surrounding the center cell. For a CDMA system, a cell layout of nineteen tri-sectored cells with hexagonal sectors will cause two tiers of co-channel interference to the center cell. Gudmundson [27] discusses microcell cell planning in Manhattan environment. The city is modeled as a chessboard with squares representing blocks and streets being located between the blocks. Different cell plans described in Section 1.3.2 can be used to simulate a microcellular environment.

Propagation Model. The performance of wireless communication systems depends on the mobile radio channel significantly. The radio wave propagates through the mobile radio channel through different mechanisms such as reflection, diffraction, and scattering. Propagation models predict the average signal strength and its variability at a given distance from the transmitter. Different propagation models exist for outdoor and indoor propagation and for different types of environments (such as urban or rural) [1]. Macrocells and microcells have different propagation characteristics. Harley [101] presents signal attenuation measurements for microcells and shows that the conventional propagation models (e.g., Hata and Okumura models) are not valid for a microcell environment. The 900 MHz and 1.8 GHz signal attenuation measurements were carried out for BS antenna heights ranging from 5 m to 20 m and an MS antenna height of 1.5m in Melbourne, Australia. The main features of the models discussed here have been experimentally validated in the literature. For example, Berg [102] suggests path loss, large-scale fading, and small-scale fading models for a microcellular system based on actual measurements. Loo [103] describes computer models of Rayleigh, Rician, log-normal, and land mobile satellite fading channels based on processing of a white Gaussian random process. The *propagation model* usually consists of a *path-loss model*, a *large-scale fading model*, and a *small-scale fading model*.

- **Path-Loss Model**. In macrocells, the path loss model is used for several aspects of cell planning such as BS placement, cell sizing, and frequency reuse [15]. The path loss models of Hata and Okumura can be used for macrocells. Microcells have different models for LOS and NLOS propagation. For an *LOS propagation*, two frequently used models are a *flat earth model* and a *two slope model*. In the flat-earth model, one direct ray and another reflected ray (with 180 degree phase shift) contribute to the total received E-field. In Berg [102], an empirical path-loss model, a two slope model, is suggested. The path-loss increases with a certain slope to a threshold distance, called a *breakpoint,* and then increases with a higher slope. In reality, wave propagation in microcells is complicated and consists of reflections and diffractions in addition to free space propagation. However, the main features of path-loss can still be described by these empirical models. For certain parameter settings, the two slope path-loss model approaches the flat earth model. For an *NLOS propagation*, a LOS propagation is assumed to the street corner. After the corner, propagation path-loss is calculated by placing an imaginary transmitter at the corner with the transmit power equaled to the power received at the corner from the LOS BS.

- **Large-Scale Fading Model**. According to [102], the distribution of the large-scale fading component is close to a log-normal distribution for a majority of LOS and NLOS streets. The distribution is actually a truncated log-normally distributed variation. In simulations, the variation should not be greater than $\pm 3\sigma$. For the measurements obtained in Berg [102], the average value of σ was found to be 4 dB for LOS streets and 3.5 dB for NLOS streets. Gudmundson [66] proposes an exponential autocorrelation model for shadow fading. The results show that the model fit is good for moderate and large cells; the predictions are less accurate for microcells due to multipaths.

- **Small-Scale Fading Model**. Small-scale or short term fading is usually modeled as a Rician distribution where parameter K (Rice factor) varies with distance. When $K = 0$, the variation is Rayleigh fading. Berg [102] suggests a small-scale fading model in terms of polynomials based on the

Rician distribution. Small-scale fading is neglected since it gets averaged out due to short correlation distance relative to that of shadow fading.

Traffic Model. Traffic can be assumed to be uniform for macrocells. However, road structures need to be considered for microcells, and traffic can be allowed only along the streets. The new call arrival process is usually modeled as an independent Poisson process. The new call durations are independent exponential random variables with a certain mean.

Mobility Model. The MSs have different velocities following a truncated Gaussian distribution. The *MS velocity* is typically assumed to be constant throughout the call.

Specific Simulation Models. A brief account of widely used simulation models is given here.

Several references [12, 51, 62, 63, 88, 93] use a two-BS model that is simple and widely used for evaluating signal strength based algorithms. This model is suitable for small macrocells and LOS handoffs in microcells. In this model, an MS travels from one BS to another in a straight line at a constant velocity. The path loss is calculated using a single slope formula, and shadow fading is assumed to be log-normal with an exponential correlation function.

A model used in [52, 53] has a four-cell neighborhood. The MS travels from one BS to another in a straight line with a constant velocity. The model assumes that there is no power control, and all BSs transmit at the same power level. The path loss is calculated using Hata's model, and shadow fading is log-normally distributed. A three-cell neighborhood instead of four-cell neighborhood is used in [64] as in [52, 53].

Two routes of an MS in a cluster of seven cells are considered in [11]. The first route is from one BS to another in which the MS crosses cell borders such that it is inside the overlapping region for a minimum duration of time. This route gives insight into the behavior of the handoff algorithm in the handoff area. The second route is from one BS to another in which the MS is in the overlapping region most of the time. This second route is more likely to have handoff complications than the first route. The four-cell model of [52] can be easily modified to create these two MS routes by adjusting the cell radii.

Chuah [56] uses an SIR-based model that can be used for integrated dynamic resource management tasks. Twenty BSs are uniformly spaced on a ring. The traffic model and the mobility models used in [56] are the same ones described earlier. The new calls are uniformly distributed throughout the ring.

A model suitable for evaluating LOS and NLOS handoffs in a microcellular environment is used in [60]. The LOS and NLOS propagation models are similar to the ones described earlier. The log-normal shadow fading with exponential correlation function for large-scale fading and Rician fading model for small-scale fading are used. The model of [60] is suitable for a microcellular environment. Two NLOS paths are considered, which give insight into the behavior of handoff algorithms when there are multiple street crossings. The effect of C/I is studied in [60] for a particular cell plan. A worst case scenario, e.g., C/I of 12 dB, is used to account for interference. Austin [60] also studies the C/I distribution for the MS and the BS.

A comprehensive model for a microcellular system is presented in [75]. This reference considers a Manhattan-like structure and places a BS at every other corner. At every street crossing, an MS either goes straight or turns with a given probability. The model is formed into a torus-like structure to avoid edge effects. The LOS propagation model is taken from [101]. For the NLOS model, it is assumed that buildings are infinitely tall, and there is a fixed loss of 20 dB every diffraction street corner. Shadow fading is not considered, but small-scale fading is modeled as Rayleigh fading.

A comprehensive simulation model suitable for macrocellular and microcellular environments is described in [13, 104, 105]. The conventional macrocellular environment is modeled by a 49-cell toroidal structure that has seven-cell clusters with 1 km radius cells [106]. The microcellular system has half-square cells with 100 m block size. The simulation model for a microcell system considers both the transmission and traffic characteristics. Such combined analysis of transmission and traffic characteristics provides a more realistic scenario for performance evaluation of a cellular system.

Rosenburg [20] gives a brief account of the simulation model, called M2 simulation, developed at AT&T; this model includes the effects of propagation, traffic, and system configuration.

The model of [73] is suitable for evaluating handoff performance in a mixed cell environment. An urban Manhattan-like environment is simulated in which a cluster consists of four microcells. Four clusters cover the service area with a macrocell overlaying the microcells. User mobility has been modeled as Gaussian with the mean value varying with the distance from the starting position of the MS. A sharp linear velocity decrease is adopted before turning, and a linear increase has been considered after the corner until the previous velocity is restored. The path loss is calculated using the two-slope model; second and fourth powers are used. The street corner is simulated by a 4 dB/m linear decrease from the street corner, lasting up to 20 m. Afterward, an NLOS propagation is assumed. Large-scale fading is simulated by uncorrelated log-normal distribution. New calls follow Poisson distribution and are uniformly distributed along the streets.

The performance of an SHO algorithm suitable for a CDMA system is analyzed in [107]. Exploitation of diversity in the cell overlap region provides better handoff performance but requires additional resources. A compromise between diversity usage and resource utilization is analyzed. The handoff performance is quantified by the performance measures such as active set updates, number of BSs involved in SHO, and outage probability. An SHO algorithm is modeled on conventional handoff algorithms. When a user is in communication with both the BSs, the user is in SHO and is said to be in the active set. A balance between the number of users in the active set and the number of active set updates is required. When a user's signal crosses a threshold, the user is added to the set. The user is removed from the set when the MS is below another threshold for a certain period of time (controlled by a timer). This timer reduces the number of active set updates significantly while increasing the average size of the set slightly.

Simmonds [85] studies the application of SHO in wideband direct sequence CDMA (DS-CDMA) systems. Propagation aspects of SHO are also presented.

Simulation results on soft/softer handoff in CDMA are presented in [87]. The effects of SHO and propagation factors including log-linear path loss and log-normal shadowing are considered. The simulation model and results on soft/softer handoff statistics for different thresholds and propagation environments are presented.

3.2.3 Emulation Approach

The emulation approach uses a software simulator consisting of a handoff algorithm to process measured variables, such as RSS and BER. Actual propagation measurement based simulation has the advantage of giving better insight into the behavior of the radio channels, and it gives more accurate data. The main disadvantages to this approach are periodic measurement requirements and inadequacy for comparison of different handoff algorithms on the same platform.

Chia [26] uses measured data in handoff simulation (the measured data was obtained by conducting 1700 MHz experiments in an urban environment in southern England). The path loss was found to follow a two slope formula with different slopes for different locations. The short term fading was found to be Rician with Rice factors varying from ten to zero depending upon the distance between the MS and the BS. It was found that the optimal handoff threshold level was different for different cites [26].

Kinoshita [67] introduces an indoor propagation simulator. The indoor simulator models trace thirteen rays over a cross-corridor and exhibit good agreement with the experiments of 950 MHz propagation with multipath fading.

Kanai [108] describes an experimental digital cellular system that consists of a Private Branch Exchange (PBX) based MSC, three BSs, two MSs, and a radio channel simulator. Experimental results indicate that the handoff decision can be made within a second, and the handoff procedure works well under typical microcell propagation conditions.

3.2.4 A Macrocellular Simulation Model

The simulation model described here is a modified version of the model used in [52] and [53].

Cell Model and Mobility Model. The cell model consists of a neighborhood of four cells as shown in Figure 3.3.

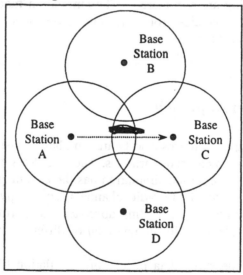

Figure 3.3: Four BS Neighborhood Cell Model.

Cell radius is 10 km, and the maximum overlap between cell A and C is 2 km. The BS Effective Isotropic Radiated Power (EIRP) is 100 W. The MS travels from one BS to another (BS A to BS C) at a constant speed, e.g., 65 mph, and such a journey is repeated 500 times. The MS velocity remains constant throughout the journey.

Propagation Model. Small-scale fading is neglected since it is averaged out due to short correlation distance relative to that of shadow fading. Thus, only path loss model and large-scale fading model are considered.

Hata's model is used to calculate path-loss. A set of equations used in the simulation model is taken from [1]. Path-loss is given by

$$L_{50(urban)} = 69.55 + 26.16\,log(f_c) - 13.82\,log(h_{te})$$
$$-a_{h_{re}} + (44.9 - 6.55\,log(h_{te}))log(d) \qquad (3.1)$$

where f_c is the carrier frequency in MHz, h_{te} is the effective BS (or transmitter) antenna height in m, h_{re} is the effective MS (or receiver) antenna height in m, d is the transmitter-receiver (T-R) separation distance in km, and $a_{h_{re}}$ is the correction factor for different sizes of the coverage area. In simulations, the following values are used: $f_c = 900$ MHz, $h_{te} = 30$ m, and $h_{re} = 3$ m.

For a medium size city,

$$a_{h_{re}} = (1.1\,log(f_c) - 0.7)h_{re} - (1.56\,log(f_c) - 0.8). \qquad (3.2)$$

In general, the path loss models of Hata and Okumura can be used for macrocells while different models are used for LOS and NLOS propagation in microcells.

According to [102], the distribution of the large-scale fading component is close to a log-normal distribution. Gudmundson [66] proposes an exponential autocorrelation model for shadow fading for macrocells and microcells that has been experimentally validated through measurements. The correlation at 100 m was found to be 0.82 for a large cell in suburban environment. The correlation was 0.3 at 10 m in a microcell. It was found that the RSS measurements are highly correlated in macrocells and less correlated in microcells. *The generation of a correlated shadow fading process from uncorrelated Gaussian samples is described next.* It is assumed that a sequence **n** is an uncorrelated process with samples that have 0 mean and σ_2 standard deviation. Assume that the sequence **n** needs to be converted into a sequence **x** with 0 mean, σ_1 standard deviation, and an exponential autocorrelation function given by

$$\rho(d) = E[x(D)x(D+d)] = \sigma_1^2\,exp\left(-\frac{d}{d_0}\right) \qquad (3.3)$$

where d is the distance separating two samples and d_0 is a parameter that can be used to specify correlation at a particular distance. For example, for a normalized autocorrelation of 0.82 at a distance of 100 m, $d_0 = 500$ m. In other words, $exp(-d/d_0) = exp(-100/500) = 0.82$.

Let x be the required correlated shadow fading process and d_s be the sampling distance. Then,

$$x(D+d_s) = \alpha x(D) + n(D).\tag{3.4}$$

The parameter α needs to be determined. The autocorrelation is

$$E[x(D+d_s)x(D)] = E[x(D)(\alpha x(D)+n(D))].\tag{3.5}$$

$$E[x(D+d_s)x(D)] = \alpha E[x(D)^2] + E[n(D)x(D)].\tag{3.6}$$

$$E[x(D+d_s)x(D)] = \alpha \sigma_1^2.\tag{3.7}$$

(since $n(D)$ and $x(D)$ are independent and 0-mean processes).
    Equating Eq. 3.3 and Eq. 3.7,

$$E[x(D+d_s)x(D)] = \rho(d) = \sigma_1^2 \, exp(-d/d_0) = \alpha \sigma_1^2.\tag{3.8}$$

Hence,

$$\alpha = exp(-d/d_0).\tag{3.9}$$

Now,

$$n(D) = x(D+d_s) - \alpha x(D).\tag{3.10}$$

$$n^2(D) = x^2(D+d_s) - 2x(D+d_s)\alpha x(D) + \alpha^2 x^2(D).\tag{3.11}$$

$$E[n^2(D)] = E[x^2(D+d_s)] - 2\alpha E[x(D+d_s)x(D)] + \alpha^2 x^2(D).\tag{3.12}$$

$$\sigma_2^2 = \sigma_1^2 - 2\alpha \sigma_1^2 \alpha + \alpha^2 \sigma_1^2.\tag{3.13}$$

$$\sigma_2^2 = \sigma_1^2 - \alpha^2 \sigma_1^2.\tag{3.14}$$

$$\sigma_2^2 = \sigma_1^2(1-\alpha^2)\tag{3.15}$$

This derivation is in conformity with the results presented in [57] and indicates that a 0-mean uncorrelated process can be used to create a 0 mean correlated process with desired variance and autocorrelation.

Traffic Model. Traffic distribution is assumed to be uniform in the four cells under consideration.

The basic simulation model used in [52] was modified to model user mobility, traffic, and interference. Traffic is quantified by the number of calls in a cell. The number of calls in the four BSs is changed uniformly between zero and 62, the maximum number of trunked channels, every seventy-five simulation steps (15% of the total simulation runs). Interference is modeled by considering a subset of the co-channel BSs, located at a distance of $D = \sqrt{3N} R$ from the center of the cell. Here, N is the cluster size, i.e., number of cells in a cluster, and R is the cell radius. The actual number of such interfering BSs is selected uniformly between zero and six every 75 simulation runs. Only the first tier of interferers is considered. Note that traffic and interference have been modeled this way to get a preliminary indication of the traffic and interference related performance of the handoff algorithm. The simulation model used here provides several important performance metrics that may be obscured in other models. The basic simulation model possesses several features. The model is simple and includes a more realistic scenario in which an MS has a four base station neighborhood and travels from one BS to another. A two base station neighborhood is a less realistic scenario. Also, this model is applicable to a macrocellular environment where BS antennas are tens of meters high and the BS transmits several Watts of power. This model can be used for a microcellular environment after making proper modifications in the propagation model. Furthermore, this model allows a detailed study of the handoff process in a handoff region. In particular, this model allows an in-depth analysis of the behavior of a handoff algorithm in the cell overlap region, and it allows elegant evaluation of certain significant handoff-related system performance metrics. The effect of different cell radii and different size cell overlap regions on the performance of a handoff algorithm can be investigated using this model.

3.2.5 A Microcellular Simulation Model

The simulation model has the following salient features.

Cell Model. A four BS neighborhood shown in Figure 3.4 is used.

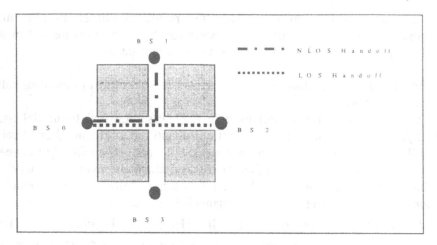

Figure 3.4: Generic Handoff Scenarios in a Microcellular System.

The BS transmit power is one Watt. The cell radius is $R = 250$ m (the same as the city block size in Figure 3.4).

Propagation Model. Microcells have different models for LOS and NLOS propagation. Here, a model similar to the one proposed in [102] is used. This is the two slope model, an empirical path loss in which the path loss increases with a certain slope to the breakpoint and then increases with a higher slope. Mathematically, RSS at a distance d from an LOS BS is given by

$$RSS(d) = P_t - (10a\,log(d) + 10b\,log(d\,/\,g)) + s(d) \qquad (3.16)$$

where P_t is the BS transmit power in Watts, d is the distance in meters of the MS from the LOS BS, and $s(d)$ is correlated shadow fading sample. The parameters a, b, and g define the path loss ($a = 2$, $b = 2$, and $g = 150$). The corner effect begins at 255 m from BS 0.

For an *NLOS propagation*, an LOS propagation is assumed up to the street corner. After the corner, propagation path loss is calculated by placing an imaginary transmitter at the corner with the transmit power equal to the power received at the corner from the LOS BS to model the diffraction effect. RSS at a distance $(d+R)$ from the NLOS BS is given by Eq. 3.16, where P_t is the power in Watts received at the intersection from the NLOS power, d is the distance of the MS from the intersection (in m), $(d+R)$ is the distance in meters of the MS from the NLOS BS, R is the distance in meters of the intersection from the NLOS BS, and $s(d)$ is the correlated shadow fading sample. Other parameters are the same as earlier. Note that P_t is calculated using the path loss formula for the LOS case (Eq. 3.16).

Correlated log-normal shadow fading is used to model large-scale fading. The log-normal shadow fading deviation is 7 dB (medium intensity). A normal range for shadow fading standard deviation for microcells is from 5 dB to 9 dB (8 dB to 14 dB for macrocells). An exponential autocorrelation model proposed in [66] is used (see Eq. 3.3). However, a correlation distance of 8.5 m is used, which gives a correlation of 0.3 at a distance of 10 m. In other words, $d_0 = 8.5$ m, and $exp(-d/d_0) = exp(-10/8.5) = 0.3$. Small-scale fading is neglected since it is averaged out due to short correlation distance relative to that of shadow fading.

Traffic, Mobility, and Interference Model. The MS travels at a constant speed from BS 0 to BS 2 during an LOS handoff scenario and from BS 0 to BS 1 during an NLOS handoff scenario. The maximum speed is 35 mph (or 15.64 m/sec), the average speed is 25 mph (or 11.18 m/sec), and the minimum speed is 15 mph (or 6.71 m/sec). There are a maximum of four LOS interferers. The maximum number of channels per BS is twenty. The number of ongoing calls in a cell and the number of interferers are chosen randomly every 10% of the total simulation time during the travel of the MS from one BS to another.

Measurement Sampling and Averaging. The measurement sample time is 0.1 sec, and the averaging distance is 6.3 m, which is a distance of 40λ. Sample averaging is used to obtain velocity adaptive averaging.

3.2.6 An Overlay Simulation Model

The simulation model proposed here allows evaluation of a handoff algorithm in generic handoff scenarios in an overlay system. This model gives details of critical performance metrics that quantify performance of significant aspects of overlay handoff. The model also gives an idea of traffic and interference related system performance. Salient features of the proposed simulation model are discussed next.

Cell Model. Figure 3.5 shows the cell layout, which consists of a cluster of seven macrocells, with each macrocell overlaying a cluster of four microcells.

Propagation Model. The propagation model used for the macrocell is the same as the model described in Section 3.2.4. For the microcell, the LOS propagation model discussed in Section 3.2.5 is used. Furthermore, small-scale fading is neglected since it is averaged out due to short correlation distance relative to that of shadow fading.

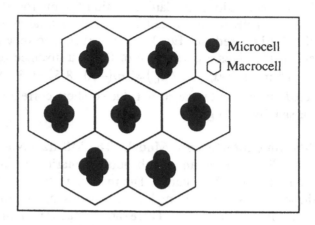

Figure 3.5: Cell Layout for an Overlay System.

Traffic, Mobility, and Interference Model. The new call durations are independent exponential random variables with 120 sec mean. Traffic is assumed to be uniform. The new call arrival process is modeled as an independent Poisson process with a certain mean arrival rate. The mean call arrival rate is given by $\lambda = Tr_{load} MB\mu$, where λ is the mean call arrival rate, Tr_{load} is the normalized traffic load (0 to 1), M is the number of channels per base station, B is the number of base stations in the cell, and $\mu = 1/cd$ where cd is the call duration in seconds. The MSs have different velocities following a truncated Gaussian distribution. The MS velocity is typically assumed to be constant throughout the call. The maximum speed is 70 mph, the average speed is 45 mph, and the minimum speed is 20 mph. There are a maximum of six co-channel interferers. The number of co-channel interferers is chosen randomly every 1% of the total simulation time for every user in the system.

Measurement Sampling and Averaging. The measurement sample time is 0.25 sec, and the averaging distance is 260 m for macrocells and 50 m for microcells. Sample averaging is used to obtain velocity adaptive averaging.

Initial Cell Selection. When a new call arrives, the call is assigned to the nearest macrocell if the user velocity is greater than the velocity threshold V_{th} or if no microcell BS can provide sufficient RSS to the MS.

3.2.7 A Soft Handoff Simulation Model

The simulation model proposed here allows evaluation of a handoff algorithm in generic handoff scenarios, such as the absence of soft handoff, two-way soft handoff, and three-way soft handoff. This model details critical performance metrics that quantify the performance of significant aspects of soft handoff such as diversity usage, resource utilization, and network load. The model also gives a preliminary idea of traffic and mobility related system performance. Salient features of the proposed simulation model are discussed next.

Cell Model. The fourteen BS neighborhood shown in Figure 3.6 is used. The BS transmit power is one Watt. The cell radius is $R = 3$ km.

Propagation Model. This model is the same as the macrocell propagation model of Section 3.2.4.

Traffic, Mobility, and Interference Model. The MS travels at a constant speed from BS 5 to BS 10. The maximum speed is 85 mph, the average speed is 65 mph, and the minimum speed is 45 mph. There is a maximum of six cochannel interferers. The maximum number of channels per BS is sixty. The number of ongoing calls in a cell and the number of interferers are chosen randomly every 10% of the total simulation time during the travel of the MS from one BS to another. Such interference modeling attempts to mimic loading variations in other cells.

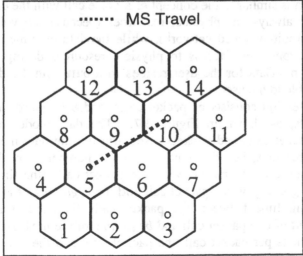

Figure 3.6: Soft Handoff Cell Layout.

Measurement Sampling and Averaging. The measurement sample time is 0.5 sec, and the averaging distance is 260 m. Sample averaging is used to obtain velocity adaptive averaging.

3.3   AN EVALUATION MECHANISM FOR RADIO RESOURCE MANAGEMENT

Since RRM in emerging systems (see Chapter 11 for details) must consider data services in addition to the voice service, ability to model data traffic becomes important. Hence, data traffic models are summarized first. A simulation framework for the analysis of RRM performance of a system is provided next.

For accurate analysis of a system carrying data traffic, modeling of data traffic characteristics is important. Models for services such as telnet, World Wide Web (WWW) and WAP (Wireless Access Protocol), file transfer protocol (ftp), e-mail, and fax are summarized here. These models are source models and describe how data is generated at the source. The model consists of three hierarchical layers of a session, packet call, and packet. The duration of the data session is the time period during which the user is exchanging information with the data network. The user may or may not have physical or radio resources throughout the session duration in a packet data network. When data is generated within a session, the user requests radio resources if the user does not already have the resources to exchange data. The concept of the data session is similar to the concept of a voice call with the exception that the voice user always has physical resources (because the voice users are served by a circuit-switched network), while the data user may not have an exclusive and continuous access to physical resources during parts of the session duration. Data for the user resides in a buffer until radio resources become available to the user.

The data session consists of packet calls, and each packet call contains several packets, as shown in Figure 3.7. The data model describes the behavior of packet calls and packets. Specifically, the data model provides analytical expressions, i.e., distributions, of the number of packet calls per session, the inter-arrival time between the packet calls (the time difference between the *beginning* of one packet call and the *beginning* of another packet call) or reading time between the packet calls (i.e., the time difference between the *end* of one packet call and *beginning* of the next packet call), the number of packets per packet call, the packet size, and the inter-arrival time between the packets.

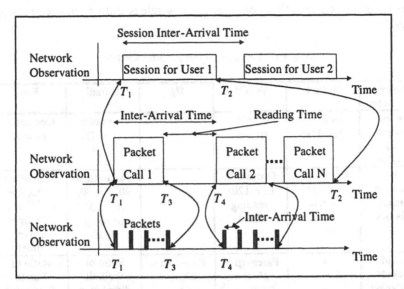

Figure 3.7: Components of a Data Traffic Model.

Different services can be assumed to conform to the structure illustrated in Figure 3.7. A telnet session consists of short interactive messages, and the message is represented by a packet call. A WWW session is the downloading of several web pages, i.e., packet calls, with a long reading time between two successive documents. An ftp session involves reception of a large file, and the packet call corresponds to this file. An e-mail session is characterized by the downloading of several e-mail messages (represented by packet calls) with some reading time between two messages. A fax session consists of reception of several pages, i.e., packet calls, with some inactivity period between the fax pages.

Table 3.1 summarizes model parameters for telnet, WWW, ftp, e-mail, and fax. In Table 3.1, μ indicates mean. The source data rate (SDR) can be changed by changing the packet inter-arrival time. A constant SDR simplifies analysis. In practice, Transmission Control Protocol/Internet Protocol (TCP/IP) adjusts the SDR based on how the RRM processes data. The models for telnet, WWW, and ftp are based on the IS-2000 and UMTS standard contributions [109] [110], while the e-mail and fax models are based on the work reported in [111], [112], and [113]. A major difference between the IS-2000 models and the UMTS models is that the IS-2000 models fix the packet size at 480 bytes and have a Pareto distributed number of packets per packet call, while the UMTS models have a Pareto distributed packet size and a geometrically distributed number of packets per packet call. Several characteristics of these models are based on observations of real traffic on landline data networks. The data models summarized in Table 3.1 can be refined based on observations of data after wireless data systems are widely deployed.

Table 3.1. Summary of Data Traffic Models.

Model Parameter	Telnet	WWW	ftp	E-mail	Fax
No. of Packet Calls Per Session	Geometric ($\mu = 114$)	Geometric ($\mu = 5$)	1	Geometric ($\mu = 2$)	Geometric ($\mu = 3$)
Time Between Packet Calls (sec)	Geometric ($\mu = 1$) (inter-arrival time)	Geometric ($\mu = 120$) (reading time)	-	Pareto ($\mu = 90$, $K = 30$, $\alpha = 1.5$) (reading time)	Weibull ($\mu = 30$, $A = 1/e^{3.52}$, $B = 2.0$) (inter-arrival time)
No. of Packets per Packet Call	1	Pareto ($\mu = 25$, $K = 2.27$, $\alpha = 1.1$)	Pareto ($\mu = 62$, $K = 5.64$, $\alpha = 1.1$)	Ratio of Weibull distributed packet call size ($\mu = 15$, $A = 1/e^9$, $B = 2.04$) and the packet size	Ratio of Pareto distributed packet call size ($\mu = 37$, $K = 1600$, $\alpha = 1.1$) and the packet size
Packet Size (bytes)	Geometric ($\mu = 90$)	480	480	480	480
Inter-Arrival Time Between Packets (sec)	-	Geometric ($\mu = 0.067$ for SDR of 57.6 kbps)	Geometric ($\mu = 0.067$ for SDR of 57.6 kbps)	Geometric ($\mu = 0.067$ for SDR of 57.6 kbps)	Geometric ($\mu = 0.27$ for SDR of 14.4 kbps)

The model parameters suggested in Table 3.1 can be used for forward link behavior of all services and forward link and reverse link behavior of telnet. To model reverse link behavior of WWW, ftp, e-mail, and fax traffic, a packet size of 90 bytes can be used (instead of 480 bytes) to emphasize that little data (mostly acknowledgments and requests) is transferred in reverse link compared to the forward link.

Distributions of random variables useful in simulating the models outlined in Table 3.1 are given here.

The CDF of an *exponential* (to approximate *geometric distribution*) variable is

$$y = F(X) = 1 - e^{-\alpha x}. \qquad (3.17)$$

The mean of the exponential distribution is $\left(\dfrac{1}{\alpha}\right)$.

Hence,
$$x = \frac{-log_e(1-y)}{\alpha}. \qquad (3.18)$$

If a uniformly distributed random variable y (between 0 and 1) is used in Eq. 3.18, x becomes a random variable that follows exponential distribution. Since geometric distribution is discrete exponential distribution, conversion of x to the closest integer can provide a variable that (approximately) follows geometric distribution.

The CDF of a Pareto distribution is

$$y = F(x) = 1 - \left(\frac{k}{x}\right)^{\alpha} \qquad (3.19)$$

where k is the minimum value of x. The mean of the normal Pareto distribution, i.e., without cut-off, is given by

$$\mu = \frac{k\alpha}{\alpha - 1}, \; \alpha > 1. \qquad (3.20)$$

If the maximum value of x is limited to be m, the mean of the Pareto distribution with cut-off becomes

$$\mu_c = \frac{\alpha k - m\left(\dfrac{k}{m}\right)^{\alpha}}{\alpha - 1}, \; \alpha > 1 \qquad (3.21)$$

From Eq. 3.19, x is found to be

$$x = \frac{k}{e^{\frac{log_e(1-y)}{\alpha}}}. \qquad (3.22)$$

If a uniformly distributed random variable y (between 0 and 1) is input into Eq. 3.22, Pareto distributed x is obtained.

The CDF of a *Weibull* distribution can be represented by

$$y = F(x) = 1 - e^{-(Ax)^B}. \qquad (3.23)$$

The mean of the Weibull distribution is

$$\mu = \frac{1}{A} \Gamma\left(1 + \frac{1}{B}\right) \qquad (3.24)$$

where $\Gamma(.)$ is a gamma function ($\Gamma(1.5) \approx 0.88$). Using Eq. 3.23,

$$x = \frac{1}{A}\left(-log_e(1-y)\right)^{\frac{1}{B}}. \qquad (3.25)$$

If a uniformly distributed random variable y (between 0 and 1) is placed in Eq. 3.25, Pareto distributed x is obtained.

Note that the random variables generated from some distributions, e.g., Pareto without cut-off, can be very large, and, they should be constrained by suitable limits. During the data traffic modeling process in a simulator, a new event, e.g., beginning of a new packet call, may occur before an old event, e.g., packet generation corresponding to previous packet call, is over. One solution to such a problem is to delay the new event until the old event is over.

There has been a growing interest in WAP. The WAP is designed to provide Internet access to wireless devices. In the WAP environment, wireless devices such as cellular phones access the Internet through a WAP gateway. The gateway translates requests from the WAP protocol stack to the WWW protocol stack and vice versa [114]. The gateway reduces the amount of data being transmitted to the wireless device through encoding. Similarities between the WWW and the WAP include self-similar traffic, i.e., the traffic that displays similar characteristics regardless of the length of observation period, and daily/weekly periodicity. The difference is that the WAP has shorter browser sessions and smaller packet sizes compared to the WWW. Thus, the WWW model summarized in Table 3.1 can potentially be employed to model the WAP by using smaller packet size, fewer packets per packet call, shorter reading time, and lower source data rate. A detailed comparison of the WWW and the WAP can be found in [114].

An RRM algorithm can be evaluated using the generic procedure outlined below. Since major third generation standards, IS-2000 and UMTS, are based on CDMA technology, *a framework suitable for CDMA systems is suggested here*. RRM for other multiple access technologies can be evaluated by customizing the suggested framework. The framework consists of four basic steps, *traffic generation, processing of session arrivals and processing of resource requests within sessions, radio interface modeling,* and *calculation of performance measures.* Traffic in a single sector is considered, and the impact of other cells, e.g., handoff and out-of-cell interference, is modeled explicitly in the traffic generation and implicitly in the radio interface modeling. As mentioned in Chapter 1, the forward link is expected to be the performance bottleneck for third generation systems based on the experience with deployed CDMA systems. Hence, the focus is on the forward link RRM here.

The main objective of the RRM simulator is to quantify the system capacity, e.g., the average number of simultaneous users, and the data throughput (total number of bits per second) for a given set of performance goals, e.g., 1% call blocking probability and short average packet delay for most packets. A major input to the RRM simulator is the average number of simultaneous users, N, which is chosen such that the desired performance goals are met. The performance of the system can be evaluated for several test cases to identify impact of various variables, e.g., impact of data penetration and impact of specific service type. For example, a test case may include 60% voice users, 20% WWW users, and 20% e-mail users. The average session duration for the assumed service distribution is found from the data traffic models. A session duration of 100 sec to 120 sec can be used for a voice session. Session arrivals are Poisson distributed. A method to achieve Poisson distributed session arrivals is to model inter-arrival time of sessions by an exponential distribution. For example, if N is 20 and the average session duration is 200 sec, the average session inter-arrival time is $200/20 = 10$ sec. Thus, an exponential distribution with a mean of 10 sec will lead to session arrivals that are Poisson distributed. Since a CDMA system utilizes soft handoff, some sessions observed in the sector are new sessions, while the remaining sessions are handoff sessions. Assume that the average sectors (or links) per user is S. If the average number of sessions modeled in the simulator is N, the average number of distinct simultaneous users is N/S, which is a good indicator of system capacity. Consider the case of ten average simultaneous sessions (a measure of system capacity in terms of carried traffic intensity) and S of 1.5. Thus, the simulator will generate $10*1.5 = 15 (= N)$ sessions to model both the new sessions, i.e., the sessions that originated in the sector under consideration, and the handoff sessions, i.e., the sessions that originated in other sector but arrives to the sector under consideration. When a session arrives, a uniform random number U is generated and compared with $1/S = 0.66$. If U is less than 0.66, the session is classified to be a new session; otherwise, it is considered to be a handoff session. Note that the average number of simultaneous sessions and the maximum number of sessions are related via Erlang-B formulation. The use of either the average number of sessions or the maximum number of sessions is one possible way of quoting the capacity.

- **Traffic Generation**. The process of traffic generation involves generation of session arrivals and generation of data for existing sessions. A new session is generated if dictated by the exponential distribution. The session is classified to be a new session or a handoff session based on the value of S using a uniform random number U as explained earlier. Next, the generated session is classified into a given type of service based on the assumed service distribution using another uniform random number V between 0 and 1. For the test case under consideration, if U is less than 0.6, the session is assumed to be voice. If U is between 0.6 and 0.8, the session is WWW, and if U is greater than 0.8, the session is e-mail. To model speech activity for a voice session, exponentially distributed speech ON-OFF periods with an average voice activity factor (i.e., the ratio of ON time period to the total (ON+OFF) time period of 0.4. An average speech ON time of 1.25 sec and average speech OFF time of 1.875 s provide an average voice activity factor of $(1.25/(1.25+1.875) = 0.4)$. The Speech ON and OFF time periods are consecutively generated. Table 3.1 can be used to model data activity for a data session.
- **Processing of Session Arrivals and Data of Existing Sessions**. A call admission control mechanism processes the requests for system access from session arrivals. Another RRM mechanism controls how resources are shared among admitted sessions. The voice information bits corresponding to speech ON times are processed immediately by the radio interface. However, the data bits are usually sent to buffers in the network. If there are insufficient (or no) resources, resource requests are made. The RRM algorithm processes such requests from existing sessions and may assign new or additional resources to users. There may be some inherent signaling delays in the system between the time when a resource request is made and the time when the resource can actually be used for transmission.
- **Radio Interface Modeling**. The data from buffer is transmitted over the air using the assigned channels of given rates. A voice user consumes power corresponding to the full rate channel, e.g., 9.6 kbps channel, during speech ON times. During speech OFF times, low rate frames (e.g., eight rate) are typically transmitted [115]. A physical layer simulator can provide a good indication of the full rate power required for a given rate channel to meet a specific frame error rate, e.g., FER of 1%. Power required for different channels with different rates can be extrapolated from the power distribution, i.e., mean and standard deviation, of the voice full rate channel. In general, the required power depends on a variety of factors, e.g., mobile speed, number of multipaths, average out-of-cell interference, and the type of handoff. Comprehensive physical layer simulations can provide the required

power as functions of all these factors. To simplify the power calculations, mean and standard deviation of the required power can be modeled as the function of the type of handoff, e.g., one-way handoff or two-way handoff. Further simplification in power calculations can be achieved by assuming a log-normally distributed power with a mean (given by the physical layer simulations that have considered all the previously mentioned factors such as mobile speed, number of multipaths, average out-of-cell interference, and the type of handoff) and an assumed standard deviation (e.g., 3 dB). A higher rate channel requires a correspondingly larger amount of power at the same FER to maintain the same signal to interference ratio. For instance, a 19.2 kbps channel may require 10% power if the power required by a 9.6 kbps channel at the same FER for the same coding technique is 5%. Influence of target FER on the required power for high rate channels is analyzed in detail in [116]. Power modeling becomes more accurate if the impact of coding, e.g., convolutional versus turbo coding, is also included. Since data services may tolerate higher FER than the voice service because of upper layer, e.g., the Medium Access Control (MAC) layer, retransmissions, a high data rate channel may be operated at higher FERs.

- **Performance Indices**. Statistics for relevant performance indices are observed or calculated if necessary.

Simulations are repeated for a predetermined time period such that steady-state results are obtained. Overall performance metrics are calculated. Section 3.1 provides details on useful performance metrics. N is adjusted (higher or lower) until the performance goals are met, e.g., 1% session blocking probability. The simulation framework outlined above generates traffic to serve N simultaneous sessions, and N serves as a measure of system capacity. Another way to generate traffic is to assume that there are M subscribers with a given distribution of services. Each subscriber generates a certain number of sessions per hour. In such a setup, M is a measure of system capacity.

3.4   SUMMARY

This chapter discusses the performance analysis of handoff and radio resource management algorithms. Different performance measures proposed in the literature for handoff and RRM are defined. Three basic evaluation mechanisms (analytical, simulation, and emulation) for handoff are briefly discussed. Simulations are more commonly used and are more versatile than the other two mechanisms. Basic constituents of the handoff simulation mechanism are explained. Detailed simulation models for various scenarios such as macrocells, microcells, overlay/underlay systems and systems employing soft handoff are described. To accurately predict the performance of a system carrying data, characteristics of data traffic are important. The models that describe the data traffic are summarized using a common framework. Such common framework can be refined when actual data traffic in emerging third generation wireless systems is observed. Data traffic models for telnet, WWW and WAP, ftp, e-mail, and fax services are summarized. Since the RRM is a complex task, an analytical approach is not suitable for the analysis. A common simulation framework for the RRM evaluation of a CDMA system is outlined.

Chapter 4

A GENERIC FUZZY LOGIC BASED HANDOFF ALGORITHM

This chapter proposes a new class of handoff algorithms that combines the attractive features of several existing algorithms and adapts the handoff parameters using fuzzy logic. Known sensitivities of handoff parameters are used to create a fuzzy logic rule base. The design procedure for a generic fuzzy logic based algorithm is outlined. Extensive simulation results for a conventional handoff algorithm (absolute and relative signal strength based algorithm) and a fuzzy logic based algorithm are presented. This chapter shows that an adaptive multicriteria fuzzy handoff algorithm performs better than a signal strength based conventional handoff algorithm. More importantly, the proposed class of algorithms allows a systematic tradeoff among different system characteristics in the dynamic cellular environment.

4.1 HANDOFF ALGORITHMS: DESIGN AND ANALYSIS ISSUES

This section discusses design and analysis procedures for a handoff algorithm. Several steps involved in the handoff algorithm design are outlined.

Figure 4.1 shows a block diagram that illustrates the design of high performance handoff. Steps involved in the handoff algorithm design and analysis are listed below.

1. **Analysis of Handoff Related System Goals**. Study handoff related cellular system goals. Analyze desirable features of a handoff algorithm, and determine the required attributes of a handoff algorithm.
2. **Determination and Preprocessing of Handoff Criteria**. Determine handoff criteria based on desired goals, system requirements, and availability of measurements. Preprocess handoff criteria before using them in a handoff algorithm. For example, some criteria, such as RSS, may need averaging.

3. **Handoff Strategy**. Process the handoff criteria using a selected strategy. Adapt the parameters of the handoff strategy by considering the performance metrics and the desired goals.
4. **Handoff Evaluation**. Evaluate the developed algorithm using a suitable simulation model.

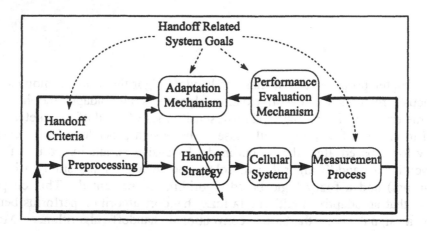

Figure 4.1: Block Diagram of a High Performance Handoff Algorithm.

This research uses the simulation approach for evaluating the performance of handoff algorithms. Details of the macrocellular simulation model used for performance evaluation of handoff algorithms are given in Section 3.2.4 of Chapter 3.

4.2 A CLASS OF FUZZY LOGIC BASED ADAPTIVE HANDOFF ALGORITHMS

4.2.1 Design Procedure

This chapter proposes a new class of adaptive handoff algorithms based on fuzzy logic, and Figure 4.2 shows a generic block diagram of this proposed class. The main idea is to combine attractive features of existing algorithms to obtain an efficient algorithm and to adapt the parameters of this efficient algorithm to the dynamic cellular environment using a fuzzy logic system.

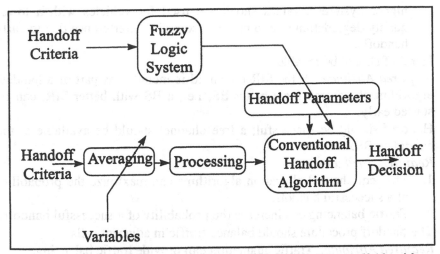

Figure 4.2: Block Diagram of Generic Fuzzy Logic Based Handoff Algorithms.

Major phases involved in the design of the proposed class of algorithms are *identification of desirable features and associated handoff algorithm attributes, selection and processing of handoff criteria, determination of the basic conventional handoff algorithm*, and *design of a fuzzy logic system*. These phases are discussed in detail next.

Identification of Desirable Features and Associated Handoff Algorithm Attributes. Two major goals of handoff algorithms are high *spectral efficiency* and high *communication quality*. Spectral efficiency is quantified by performance metrics such as call blocking probability and handoff blocking probability. Communication quality is quantified by performance metrics such as SIR and the number of dropped calls. Several desirable features of handoff algorithms are explained in Chapter 1. An efficient handoff algorithm can achieve many of these features by possessing certain attributes and making appropriate tradeoffs among various operating characteristics. The attributes of an algorithm associated with these features are discussed next.

- Handoff should be fast so the user does not experience service degradation or service interruption.
  Required Attributes.
  1. A handoff algorithm should be simple so it can be executed quickly.
  2. There should be a margin between the call drop RSS threshold and the handoff RSS threshold and between the call drop SIR threshold and the handoff SIR threshold.
  3. Higher threshold and lower hysteresis values should be used for vehicles with a higher quality degradation rate. Lower threshold and

higher hysteresis values should be used for vehicles with a lower quality degradation rate to prevent excessive interference due to early handoffs.

- Handoff should be reliable.

Required Attribute. If the SIR threshold is included as part of a handoff algorithm, the search for a better BS, i.e., a BS with better SIR, can be started early.

- Handoff should be successful; a free channel should be available at the candidate BS.

Required Attributes.
1. Efficient channel allocation algorithms can maximize the probability of a successful handoff.
2. Traffic balancing can increase the probability of a successful handoff.

- The handoff procedure should balance traffic in adjacent cells.

Required Attribute. Traffic adaptation can provide traffic balancing.

- Handoff should maintain the planned cellular borders.

Required Attributes.
1. Since RSS is an indicator of the distance between the BS and the MS, an RSS based algorithm can help preserve planned cell boundaries.
2. It should be noted that even though intrinsic traffic balancing perturbs planned cell boundaries, combined traffic and interference adaptation can achieve a systematic tradeoff between traffic balancing and resultant interference.

- The number of handoffs should be minimized.

Required Attributes.
1. Identification of a correct target cell prevents unnecessary and frequent handoffs.
2. Velocity adaptive averaging, hysteresis, and direction-biasing can help reduce the number of handoffs.

- The handoff procedure should minimize the number of call drops.

Required Attribute. A handoff algorithm should provide sufficient RSS and SIR.

- The global interference level should be minimized by the handoff procedure.

Required Attribute. Better RSS and SIR distribution allow the MS to transmit low power, reducing the overall interference.

Based on the required attributes and the study of the properties and behavior of existing algorithms, a configuration of a generic handoff algorithm, shown in Figures 4.3 and 4.4, is derived.

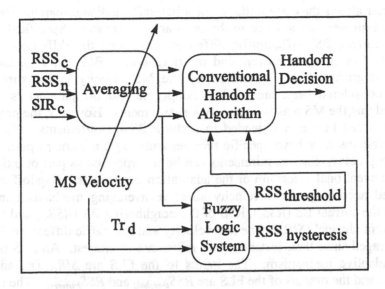

Figure 4.3: A Generic Adaptive Fuzzy Logic Based Algorithm.

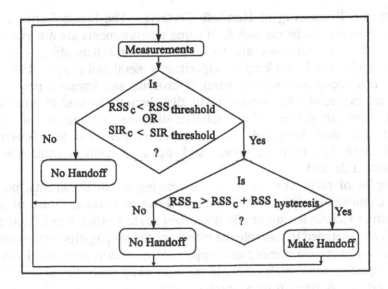

Figure 4.4: The Conventional Handoff Algorithm
for the Generic Fuzzy Logic Based Algorithm.

This configuration uses the combination of an absolute and relative RSS based algorithm and an SIR based algorithm. The RSS based algorithm has threshold ($RSS_{threshold}$) and hysteresis ($RSS_{hysteresis}$) as parameters, while the SIR based algorithm has threshold ($SIR_{threshold}$) as a parameter. The SIR threshold parameter allows the early initiation of a handoff candidate search. The RSS based parameters are adapted to the cellular environment. Specifically, SIR of the current BS, SIR_c, traffic difference, Tr_d, i.e., the difference in the number of calls in the current and the neighboring BS, $(Tr_c\text{-}Tr_n)$, and call quality degradation rate, quantified by the MS velocity in the simulation model considered here, are used to adapt the handoff parameters. It is assumed that the MS makes necessary measurements. However, the proposed algorithm can be easily extended to include BS measurements. Note that actual systems may have specific requirements, e.g., maximum permissible round-trip delay. Such requirements can be incorporated as part of either the basic conventional algorithm or the adaptation mechanism. Handoff criteria averaged according to the velocity adaptive averaging mechanism include RSS of the current BS (RSS_c), RSS of the neighboring BS (RSS_n), and SIR of the current channel (SIR_c). The MS velocity and the traffic difference Tr_d are not averaged since their instantaneous values are of interest. An FLS is used as an adaptive mechanism. The inputs to the FLS are SIR_c, Tr_d, and *MS velocity*, and the outputs of the FLS are $RSS_{threshold}$ and $RSS_{hysteresis}$. The details of the FLS are given later in this section.

Selection and Processing of Handoff Criteria. The handoff criteria and other variables need to be measured. If some measurements are not available, estimates of these variables are required. Several handoff criteria are discussed in Chapter 1. To keep the algorithm general and simple, RSS, SIR, traffic, and velocity are used as handoff criteria, and transmit power and distance are excluded. An improved SIR distribution can lead to potentially lower MS transmit power. Since relative distance measurement accuracy decreases with decreasing cell size, distance is not used as a criterion. However, both MS transmit power and distance measurements can be incorporated if desired.

Examples of preprocessing include averaging of measurements, putting different emphases on different criteria, and direction-biasing. Some of these preprocessing techniques are briefly discussed next. Certain handoff criteria, e.g., RSS, are averaged to mitigate the effects of the propagation environment. To prevent an excessive number of dropped calls, handoff requests should be processed quickly for vehicles that are moving away from the serving BS at high velocities. A fixed time averaging interval gives best performance at only one velocity. For example, for a fixed parameter handoff algorithm with a fixed time averaging window, two situations exist: (i) for high velocities, handoff delay is long, and the number of handoffs are fewer and (ii) for low velocities, the handoff delay is short and the number of handoffs is more. Also, there is a velocity between the high and low extremes that gives optimum performance for both the handoff delay and the number of handoffs. To provide similar performance, i.e., the desired tradeoff between the handoff delay and the number of handoffs, to users with different velocities, the temporal window should be adapted based on the MS velocity. Velocity adaptive averaging algorithms for a microcellular environment are proposed in [58]. These algorithms provide good performance for MSs with different velocities by adjusting the effective length of the averaging window. A velocity adaptive handoff algorithm can serve as an alternative to the umbrella cell approach to tackle high-speed users if low network delay can be achieved, which can lead to savings in the infrastructure. The temporal window length can be changed by either keeping the sampling period constant and adjusting the number of samples per window, or vice versa. For microcells, the sampling distance of 0.5λ and the averaging distance of 20λ to 40λ are sufficient according to [58]. In this chapter, the measurement sampling period T_s is 0.5 sec (as in GSM), and the number of samples per window are adjusted according to the *MS velocity*. A spatial window of 780λ, i.e., 260 m for $f_c=900$ MHz, is used. The averaging distance of 260 m was determined based on simulations. For a velocity of 65 mph (or 29.06 m/sec), eighteen samples are used, while for a velocity of 85 mph (or 38 m/sec), fourteen samples are used. This chapter assumes that velocity estimation is available. Several methods for estimating velocity are described in [58].

The averaged criteria can be further processed before their use in a handoff algorithm. For example, each criterion can be emphasized differently. Also, some BSs may be favored based on the direction in which the MS is moving; this process is called *direction-biasing*. The basic idea behind direction biasing is to encourage handoff to the BS that the MS is approaching and to discourage handoff to the BS from which the MS is receding. Direction-biased handoff algorithms that have better cell membership properties, defined as a probability of close to one throughout the call duration, are proposed in [60]. Improved cell membership properties can help preserve cell boundaries, thereby reducing the interference, and reduce the number of handoffs.

Determination of the Basic Conventional Handoff Algorithm. A conventional algorithm based on RSS and SIR was used in conjunction with the FLS. This algorithm is shown in Figure 4.4. According to the absolute and relative RSS-based part of the algorithm, handoff takes place if the following two conditions are satisfied [54]: (i) the average signal strength of the serving BS falls below an absolute threshold ($RSS_{threshold}$) and (ii) the average signal strength of the candidate BS exceeds the average signal strength of the current BS by an amount of ($RSS_{hysteresis}$). Condition (i) prevents the occurrence of handoff when the current BS can provide sufficient signal quality, while condition (ii) reduces the ping-pong effect. Beck [24] has shown that an optimum threshold achieves a narrowed handoff area (and hence reduced interference) and a low expected number of handoffs.

According to the SIR based part of the algorithm, a handoff candidate search is initiated when SIR_c drops below a threshold, $SIR_{threshold}$. An algorithm based on both RSS and BER is described in [26]. For RSS, a threshold is used for the current BS, and a hysteresis window is used for the target BS. For BER, a separate threshold is defined. The target BS can be either included or excluded from the handoff decision process. The latter scheme is used in GSM where the mobile does not know the signal quality of the target BS.

Design of the Fuzzy Logic System. A fuzzy logic rule base is created based on the known sensitivity of handoff algorithm parameters, e.g., $RSS_{threshold}$ and $RSS_{hysteresis}$, to interference and traffic. This research utilizes the Mamdani FLS described in Chapter 2. The singleton fuzzifier, the product operation fuzzy implication for fuzzy inference, and the center average defuzzifier are used. Each of the input fuzzy variables is assigned to one of the three fuzzy sets, "High," "Normal," or "Low." Each of the output variables is assigned one of the seven fuzzy sets, "Highest," "Higher," "High," "Normal," "Low," "Lower," or "Lowest." An example of the universes of discourse for the input and output fuzzy variables is shown in Figure 4.5. For example, consider the fuzzy variable SIR_c. Its universe of discourse is defined from 14 dB to 22 dB. The fuzzy set "High" for the SIR_c is defined from 18 dB to 22 dB with the maximum membership at 22 dB. Similarly, the fuzzy set "Normal" for the SIR_c is defined from 14 dB to 22 dB with the maximum membership at 18 dB, and the fuzzy set "Low" for the SIR_c is defined from 14 dB to 18 dB with the maximum membership at 14 dB. The degrees of freedom for each of the fuzzy variables are centers of the Gaussian membership functions, spreads of the membership functions, and ranges of the universes of discourse. If equal weight is given to the input fuzzy variables, the creation of the fuzzy logic rule base is facilitated. The sensitivity of the input fuzzy variable to the output of the FLS can be controlled by changing the universe of discourse. Moreover, the addition of more fuzzy sets in a given universe of discourse can give improved resolution and better sensitivity control. To keep the complexity of the fuzzy logic rule base low, the universe of discourse for each input fuzzy variable was classified into three fuzzy sets. The universe of discourse for the output fuzzy variable was divided into seven regions so that appropriate weight can be given to the different combinations of the input fuzzy sets. The universe of discourse for *MS velocity* is from 18 m/s to 40 m/s in Figure 4.5 as an example. In practice, the range should be from negative maximum expected velocity to positive maximum expected velocity. Basically, handoff should be discouraged if the MS is moving at a high velocity toward the current BS. Handoff should be encouraged if the MS is moving toward the candidate BS at a high velocity.

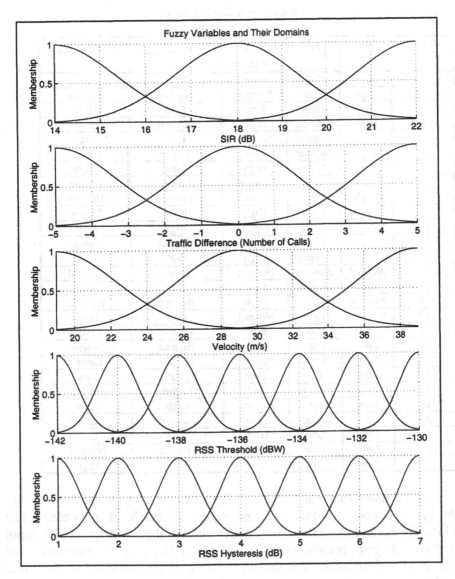

Figure 4.5: Membership Functions of Fuzzy Variables.

The fuzzy rule base is shown in Table 4.1. Assume that SIR_c is "Low," Tr_d is "High," and *MS velocity* is "High." These conditions indicate that the handoff should be *encouraged* as much as possible; this is rule number nineteen. To make the fastest handoff, $RSS_{threshold}$ is increased to the highest value, and $RSS_{hysteresis}$ is decreased to the lowest value. In practice, "Low" should refer to the high velocity toward the current BS and "High" velocity should refer to the high velocity toward the candidate BS.

Table 4.1: Fuzzy Logic Rule Base.

Rule Number	SIR_c	Tr_d	MS Velocity	$\Delta RSS_{threshold}$	$\Delta RSS_{hysteresis}$
1	High	High	High	High	Low
2	High	High	Normal	Normal	Normal
3	High	High	Low	Low	High
4	High	Normal	High	Normal	Normal
5	High	Normal	Normal	Low	High
6	High	Normal	Low	Lower	Higher
7	High	Low	High	Low	High
8	High	Low	Normal	Lower	Higher
9	High	Low	Low	Lowest	Highest
10	Normal	High	High	Higher	Lower
11	Normal	High	Normal	High	Low
12	Normal	High	Low	Normal	Normal
13	Normal	Normal	High	High	Low
14	Normal	Normal	Normal	Normal	Normal
15	Normal	Normal	Low	Low	High
16	Normal	Low	High	Normal	Normal
17	Normal	Low	Normal	Low	High
18	Normal	Low	Low	Lower	Higher
19	Low	High	High	Highest	Lowest
20	Low	High	Normal	Higher	Lower
21	Low	High	Low	High	Low
22	Low	Normal	High	Higher	Lower
23	Low	Normal	Normal	High	Low
24	Low	Normal	Low	Normal	Normal
25	Low	Low	High	High	Low
26	Low	Low	Normal	Normal	Normal
27	Low	Low	Low	Low	High

Now consider rule nine. Since SIR_c is "High," Tr_d is "Low," *and MS velocity* is "Low," handoff is *discouraged* as much as possible. The philosophy behind the other rules states that if the majority of the input variables suggest encouraging a handoff, $RSS_{threshold}$ is increased and $RSS_{hysteresis}$ is decreased. On the other hand, if the majority of the input variables suggest discouraging a handoff, $RSS_{threshold}$ is decreased, and $RSS_{hysteresis}$ is increased. The extent to which the threshold and hysteresis are changed to encourage or discourage a handoff depends upon how many variables agree on a particular direction of the threshold and hysteresis change. Resolving conflicting criteria in accordance with the global system goals is an important advantage of fuzzy logic. For example, consider rule two. "High" SIR_c discourages a handoff, while "High" Tr_d encourages a handoff. *MS velocity* is neutral. Hence, the fuzzy logic rule makes the logical decision to keep $RSS_{threshold}$ and $RSS_{hysteresis}$ values nominal.

4.2.2 Analysis of Proposed Class of Algorithms

The proposed class of fuzzy algorithms has a number of advantages over existing algorithms. Some of the significant advantages are mentioned here.

- **Use of Attractive Features of Existing Algorithms.** The proposed algorithm utilizes several attractive features of existing algorithms. For example, a combined RSS and SIR based algorithm with handoff parameters, such as threshold and hysteresis, is used as a conventional algorithm in Figure 4.2. This algorithm leads to fewer handoffs, reduced ping-pong effect, and fewer unnecessary handoffs. The proposed class of algorithms does not *replace* the existing algorithms; it *complements* the existing algorithms. The proposed technique enhances the performance of conventional algorithms by providing a robust adaptation mechanism to make appropriate tradeoffs.

- **Use of Multiple Handoff Criteria.** Multicriteria algorithms provide better performance than single criterion algorithms due to the additional flexibility and complementary nature of the criteria. For example, consider a situation in which (i) the traffic in the handoff candidate cell is low and (ii) SIR for the current cell is very high. If an adaptive single criterion handoff algorithm based on traffic is used, it would increase the handoff threshold, encouraging a handoff. An adaptive single criterion handoff algorithm based on SIR would lower the handoff threshold, discouraging a handoff. However, a multicriteria handoff algorithm can consider both measurements (traffic and SIR) and make a decision that is consistent with the global system goals.

- **Adaptation.** The proposed algorithm is adaptive to interference, traffic, and velocity. Adaptation to interference gives better RSS and SIR distribution, resulting in fewer dropped calls, better communication quality, potentially lower MS transmit power requirements, and better cell memberships. Adaptation to traffic provides traffic balancing, reducing the blocking probability of new calls and handoff calls. Adaptation to velocity (or quality degradation rate) gives good performance at different MS speeds.

- **Fuzzy Logic Attributes.** One of the advantages of fuzzy logic is simplicity. Furthermore, the knowledge about the system can be better exploited with fuzzy logic algorithms than with conventional algorithms [89]. Conflicting criteria can also be resolved using fuzzy logic.

- **Systematic Balance.** The proposed algorithm allows a systematic compromise among various characteristics by properly tuning the parameters of the fuzzy logic rule base.

Increased complexity is a disadvantage of the proposed algorithms compared to conventional algorithms. Nevertheless, inherent parallelism in an FLS partially offsets this increase in complexity. There are several ways of reducing this complexity. For example, an FLS can be represented compactly using some neural network paradigms, leading to the savings in the storage and computational requirements (see Chapter 5).

4.3   PERFORMANCE ANALYSIS OF PROPOSED AND CONVENTIONAL ALGORITHMS

The performances of the proposed and conventional handoff algorithms have been evaluated using several performance metrics, covering major aspects of handoff related system performance. These performance metrics include *CDF of RSS, CDF of SIR, CDF of traffic, average number of handoffs*, and *cross-over distance*. When an MS is connected to a BS, the RSS from the BS, the downlink SIR, and the number of calls in the BS are stored, and these stored values are used to derive CDFs of RSS, SIR, and traffic, respectively. The concept of an *operating point* is used for one aspect of performance evaluation. The operating point is defined by the (x,y) pair where x is the 50% cross-over distance and y is the average number of handoffs during a travel. The average 50% cross-over distance is the distance of the MS from the BS where the probability that the MS is connected to the BS is 0.5 and is indicative of meeting the desire to define cell boundaries. Ideally, it is desired that the number of handoffs be minimum and the cross-over distance be as close as possible to the midpoint of the travel. The CDFs of RSS and SIR can imply the call drop probability, call quality, and potential uplink transmit power requirements. For example, high values of RSS and SIR indicate that the call drop probability will be low, the call quality will be good, and the MS can transmit comparatively less power. The traffic CDF can give an idea of traffic balancing (or new call or handoff blocking probability). For example, a lower number of calls in a cell imply a high probability of successful network access since more new calls or handoff requests can be entertained by the network. A lower number of handoffs indicates a lower switching load and a shorter delay in the processing of a handoff request. The cross-over distance reflects the interference level, handoff delay, and MS power requirements. *In most of the simulation results reported in this book, portions of the CDFs of variables (e.g., lower portion for RSS and SIR and upper portion for traffic) are shown, because these portions are the regions of interest.*

4.3.1 Interference Adaptation

Figure 4.6 shows the CDF of SIR for a conventional algorithm and the fuzzy algorithm, illustrating that the CDF of the SIR for a fuzzy algorithm is to the right of the CDF of SIR for the conventional algorithm with an improvement of approximately 1.0 dB.

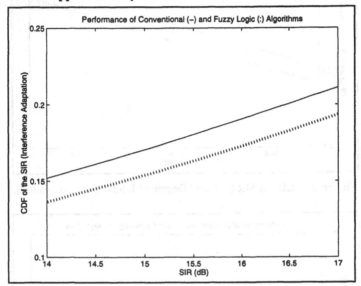

Figure 4.6: CDF of SIR ("Normal" Degree of Interference Adaptation).

Note that the traffic and velocity adaptation were not in effect when this simulation result was obtained, isolating the performance improvement due to interference adaptation and showing the function of the interference adaptation part of the fuzzy logic rule base. Also note that the improvement of SIR can be changed by tuning the fuzzy logic parameters. For example, Figure 4.7 shows the CDF of SIR when there was less improvement (0.6 dB) compared to Figure 4.6 tuning. On the other hand, Figure 4.8 shows the CDF of SIR when there was more improvement (1.5 dB) compared to the Figure 4.6 tuning. The FLS parameters can be tuned in a number of ways, and here, the definitions of the fuzzy sets "High" and "Low" for the fuzzy variables $RSS_{threshold}$ and $RSS_{hysteresis}$ were changed.

The centers of the membership functions for the fuzzy set "High" of the fuzzy variable $RSS_{threshold}$ were -130 dBm, -132 dBm, and -134 dBm for Figure 4.8, Figure 4.6, and Figure 4.7, respectively. The centers of the membership functions for the fuzzy set "Low" of the fuzzy variable $RSS_{threshold}$ were -142 dBm, -140 dBm, and -138 dBm for Figure 4.8, Figure 4.6, and Figure 4.7, respectively.

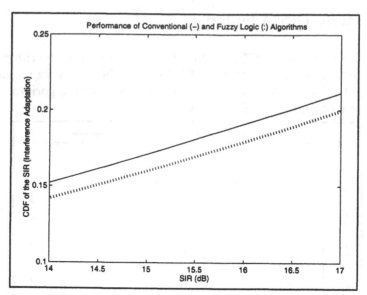

Figure 4.7: CDF of SIR ("Lower" Degree of Interference Adaptation).

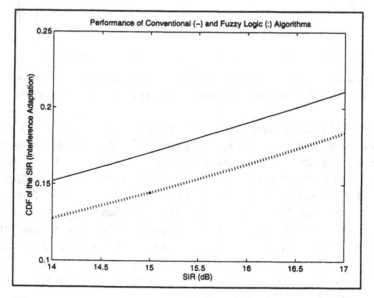

Figure 4.8: CDF of SIR ("Higher" Degree of Interference Adaptation).

The centers of the membership functions for the fuzzy set "High" of the fuzzy variable $RSS_{hysteresis}$ were 20.0 dB, 18.5 dB, and 17.0 dB for Figure 4.8, Figure 4.6, and Figure 4.7, respectively. The centers of the membership functions for the fuzzy set "Low" of the fuzzy variable $RSS_{hysteresis}$ were 12.0 dB, 13.5 dB, and 15.0 dB for Figure 4.8, Figure 4.6, and Figure 4.7, respectively. The same definitions of the "High" and "Low" fuzzy sets for the fuzzy variables $RSS_{threshold}$ and $RSS_{hysteresis}$ were used to obtain "Higher," "Normal," and "Lower" degrees of adaptation for different test cases (such as interference adaptation and traffic adaptation). The centers of the membership functions for the fuzzy sets "Normal" of the fuzzy variables $RSS_{threshold}$ and $RSS_{hysteresis}$ were chosen to be -136 dBm and 16.0 dB, respectively, in all the test cases. Assume that the call is dropped at SIR of 14.0 dB. An approximate idea about the call drop probability can be obtained by comparing the probabilities of SIR distribution at this value of SIR for the conventional and fuzzy algorithms. The probability that SIR is less than 14 dB is 0.1519 for the conventional algorithm and 0.1367 for the fuzzy algorithm with "Normal" degree of interference adaptation. Thus, there is an improvement of ((0.1519-0.1362)/0.1519) 100% = 10%. For a "Low" degree of interference adaptation, there is an improvement of ((0.1519-0.1418)/0.1519)100% = 7%, and for a "High" degree of interference adaptation, there is an improvement of ((0.1519-0.1275)/0.1519)100% = 16%. Thus, there can be a 7% to 16% improvement in call drop probability depending upon the tuning of the fuzzy logic parameters. The maximum possible improvement depends on system related constraints, e.g., the maximum permissible MS transmit power and the maximum permissible distance between an MS and a BS. Field-testing can be used to refine the parameter settings of the algorithm.

4.3.2 Traffic Adaptation

The traffic adaptation capability of the fuzzy logic system is demonstrated next using a traffic distribution. The interference and velocity adaptation were switched off during this simulation to clearly show the effect of traffic adaptation. Figure 4.9 shows that the CDF of traffic for the fuzzy algorithm is to the left of the CDF of traffic for the conventional algorithm, giving an improvement of 1.8 calls in the traffic distribution.

Also note that the improvement of traffic can be changed by tuning the fuzzy logic parameters. There is a 1.0 call improvement in the traffic distribution for the lower degree of adaptation and a 2.8 call improvement in the traffic distribution for the higher degree of adaptation.

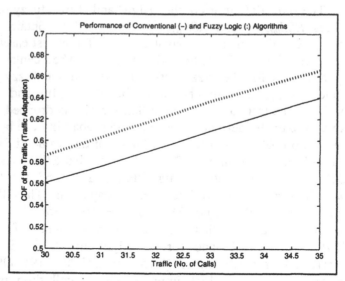

Figure 4.9: CDF of Traffic ("Normal" Degree of Adaptation).

4.3.3 Velocity Adaptation

The velocity adaptation capability of the fuzzy logic system is demonstrated next for the cross-over distance. The interference and traffic adaptation were switched off during this simulation result to show the effect of velocity adaptation. The minimum velocity is 45 mph (20 m/s), the average velocity is 65 mph (29 m/s), and the maximum velocity is 85 mph (38 m/s). The operating point for the fuzzy algorithm has a shorter cross-over distance compared to the conventional algorithm because the fuzzy algorithm tends to reduce cross-over distance, giving less interference. The improvement in the handoff delay can be determined by comparing the reduction in cross-over distance for different velocities. For the low velocity (20 m/s), there is a reduction of 60 m in cross-over distance, an improvement of 60/20 = 3.0 sec in handoff delay. For the average velocity (29 m/s), there is a reduction of 150 m in cross-over distance, i.e., an improvement of 150/29 = 5.1 sec in handoff delay. For the high velocity (38 m/s), there is a reduction of 230 m in cross-over distance, an improvement of 2350/38 = 6.1 sec in handoff delay. Thus, the improvement in handoff delay conforms to the desired goal, i.e., handoff should be faster for high velocities.

4.3.4 Combined Interference, Traffic, and Velocity Adaptation

The RSS distribution is found to be very similar for conventional and fuzzy algorithms when all the components of adaptation are in effect simultaneously. There is a marginal improvement (a fraction of a dB) with the fuzzy algorithm.

Figure 4.10 shows the SIR distribution when all the components of adaptation are in effect simultaneously. The fuzzy algorithm shows an improvement of 0.8 dB over the conventional algorithm. Note that the SIR improvement depends on which rules contribute to the overall fuzzy logic outputs. In practice, certain situations will give higher improvement than that shown here. For example, there may have been numerous occasions where conflicting requirements would have forced the FLS to keep the handoff parameters as nominal values, limiting the perceived SIR improvement. If several events indicate that the handoff parameter changes to a higher degree, more improvement in SIR would be visible.

Figure 4.11 shows traffic distribution when all the components of adaptation were in effect simultaneously, and the fuzzy algorithm shows an improvement of 1.8 calls over the conventional algorithm. As discussed earlier for SIR, certain situations in practice will give higher improvement than that shown here. For example, if there is a nonuniform traffic distribution, more improvement would be feasible.

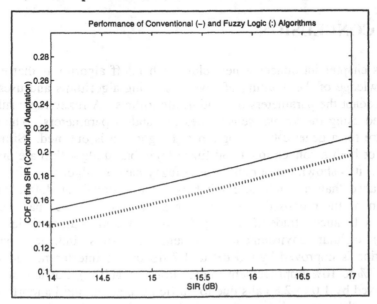

Figure 4.10: Effect of Combined Adaptation on SIR Performance.

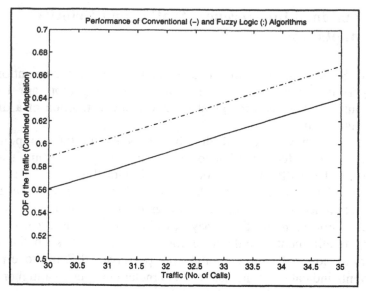

Figure 4.11: Effect of Combined Adaptation on Traffic Performance.

When all the components of adaptation work together for an average velocity of 65 mph (29 m/s), the handoff delay is reduced by 160 m or 160/29= 5.5 sec with the fuzzy algorithm.

4.4   CONCLUSION

This chapter introduces a new class of handoff algorithms that exploits the knowledge of the working of several existing algorithms and uses fuzzy logic to adapt the parameters of handoff algorithms. A fuzzy logic rule base is created using the known sensitivities of handoff parameters. The design procedure for a generic fuzzy logic based algorithm is outlined. Simulation results for both a conventional and fuzzy logic based algorithm are analyzed in detail. It is shown that a multicriteria fuzzy handoff algorithm gives better performance than a signal strength based conventional handoff algorithm. Furthermore, the proposed approach allows tuning of the FLS parameters to achieve a balanced tradeoff among different system characteristics in the dynamic cellular environment. Simulation results indicate that SIR distribution is improved by 0.5 dB to 1.7 dB due to interference adaptation (giving 7% to 16% improvement in call drop probability), traffic distribution is improved by 1.0 to 2.8 calls due to traffic adaptation, and handoff delay is reduced by three seconds to six seconds due to velocity adaptation.

Chapter 5

A NEURAL ENCODED FUZZY LOGIC ALGORITHM

This chapter proposes a new class of handoff algorithms that adapts the parameters of a handoff algorithm using a neural encoded fuzzy logic system. Known sensitivities of handoff parameters can be used to design an FLS, which can then be used to adapt the handoff parameters to obtain improved performance in a dynamic cellular environment. However, the FLS has large storage requirements and high computational complexity. This chapter proposes neural encoding of the FLS to circumvent these demands; a neural network learns how the FLS works. Several neural network paradigms such as a multilayer perceptron and a radial basis function network can be universal approximators. The input-output mapping capability and compact data representation capability of neural networks are exploited here to derive an adaptive handoff algorithm that retains the high performance of the original fuzzy logic based algorithm and that has an efficient architecture for storage and computational requirements.

5.1   INTRODUCTION TO AN ADAPTIVE HANDOFF ALGORITHM

An adaptive handoff algorithm based on neural networks is proposed in this chapter. Known sensitivities of handoff parameters can be used to create an FLS. Since neural networks can represent information compactly, good savings in storage and computational requirements can be obtained if the fuzzy logic rule base is replaced by a neural network. This chapter discusses the utilization of two neural networks, an MLP and an RBFN, to mimic the working of the FLS. These network paradigms are trained to learn the relationship among the inputs and the outputs of the fuzzy logic rule base. The trained neural networks are used to adapt the parameters of a handoff algorithm. The performance of this adaptive neural handoff algorithm is compared with the conventional handoff algorithm. The simulation results show that the proposed neural algorithm possesses a very low complexity architecture but retains the high performance of the original fuzzy logic based handoff algorithm.

The configuration of a generic handoff algorithm, shown in Figures 5.1 and 5.2, was proposed in [117] and described in Chapter 4.

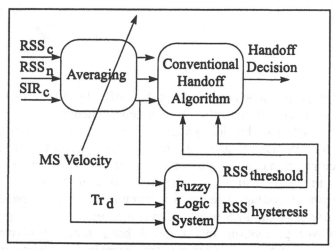

Figure 5.1: An Adaptive Fuzzy Logic Based Algorithm.

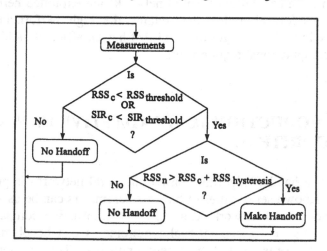

Figure 5.2: A Conventional Algorithm for a Generic Fuzzy Logic Based Handoff Algorithm.

This configuration uses the combination of an absolute and relative RSS based algorithm and an SIR based algorithm. The RSS based algorithm has threshold ($RSS_{threshold}$) and hysteresis ($RSS_{hysteresis}$) as parameters, while the SIR based algorithm has threshold ($SIR_{threshold}$) as a parameter. The SIR threshold parameter allows the initiation of the search for a better handoff candidate early on. The RSS based parameters are adapted using the FLS. It is assumed that the MS makes necessary measurements. However, the proposed algorithm can be easily extended to include BS measurements. Note that in actual systems, there may be specific requirements, e.g., maximum permissible round-trip delay. Such requirements can be incorporated as part of either the basic conventional algorithm or the adaptation mechanism. Handoff criteria that are averaged according to the velocity adaptive averaging mechanism include RSS of the current BS (RSS_c), RSS of the neighboring BS (RSS_n), and SIR of the current channel (SIR_c). The MS velocity and the traffic difference Tr_d , i.e., the difference in the number of calls in the current BS and the neighboring BS, are not averaged since their instantaneous values are of interest. An FLS is used as the adaptation mechanism. The inputs to the FLS are SIR_c, $Tr_d = Tr_c - Tr_n$ (Tr_c is the number of calls in the current BS, and Tr_n is the number of calls in the neighboring BS), and *MS velocity*. The outputs of the FLS are $RSS_{threshold}$ and $RSS_{hysteresis}$.

The inherent parallelism in FLSs allows an efficient implementation of the fuzzy logic based algorithm. However, the algorithm is still much more complex than conventional algorithms that consist of only a few binary IF-THEN rules. Moreover, as the number of inputs to the FLS increases or as the universes of discourse for the fuzzy variables are divided into more fuzzy sets, the complexity of the FLS increases even further. The complexity is two-fold, storage requirements and the number of computations to be performed every measurement sample time. A simple handoff algorithm with fewer computations and less storage requirements is desirable since it can be executed quickly and does not consume a significant amount of the available resources. The simplicity of a handoff algorithm is becoming more and more important as the user demand in cellular systems is expected to increase in the coming several years. The simplicity of the algorithm can reduce the handoff delay, the number of dropped calls, and the number of blocked calls. Moreover, the savings in computation time gives the processing unit, e.g., an MSC or a BS, an opportunity to devote time to other aspects (such as collection of measurements or intelligent resource allocation) to improve the overall system performance. Hence, it is advantageous to reduce the complexity to achieve a relatively simple algorithm with faster execution time. In this chapter, neural encoding of the FLS is proposed to simultaneously achieve the goals of high performance and reduced complexity. The storage requirements and computational savings are analyzed. The neural encoded FLS based algorithm is evaluated comprehensively, and its performance is compared with a conventional absolute and relative RSS based algorithm and the original fuzzy logic based algorithm. The chapter shows that the neural encoded FLS algorithm (NEFLSA) performs as well as the basic fuzzy logic algorithm (FLA) and that the NEFLSA is less complex than the FLA. The parameters of neural networks allow a tradeoff between the complexity and the approximation accuracy. The simulation results shown in this chapter are not intended to be the optimum results obtainable using fuzzy logic or neural networks, but rather illustrate the principle and potential of this approach. A particular configuration of fuzzy logic system designed in [117] and described in Chapter 4 was chosen as a basic FLS, and two paradigms of neural networks, MLP and RBFN, were trained to mimic the operation of this FLS.

Section 5.2 describes a procedure for using an ANN as an adaptation mechanism in place of the FLS. The performances of a conventional algorithm and the proposed neural algorithm are evaluated in Section 5.3. Finally, Section 5.4 summarizes the chapter.

5.2 APPLICATION OF NEURAL NETWORKS TO HANDOFF

An ANN can be trained to learn complex relationships among the inputs and outputs of a system. After the ANN is trained, the parameters of the ANN can be used to estimate the outputs for given inputs. Figure 5.3 shows the flowchart illustrating the training mechanism of the ANN in a supervised learning mode. A generic training procedure with its application to the FLS mapping is explained next.

1. Get the data set that contains the system inputs and desired outputs. Different possible combinations of the inputs are applied to the FLSs, and the corresponding outputs of the FLS are observed. The FLS inputs and outputs constitute the data set.

2. Determine the structure of the neural network and the associated learning algorithm. Since the problem of function approximation is a generalization problem, two suitable paradigms, MLP and RBFN, are considered here.

3. Determine the input set and the target output set for the neural network. Since the mapping between the FLS inputs and outputs is static, input and output data sets collected in Step 1 can serve as the input set and the target output set with proper scaling. It is important to scale the training data set so that the network parameters do not saturate. In general, the input data set obtained in Step 1 may need to be preprocessed for use with the ANN when the system outputs depend not only on the current inputs but also on the history of inputs and outputs.

4. Select the training parameters (such as the learning rate and number of neurons) and train the neural network using an appropriate algorithm. If the network does not perform satisfactorily, several

possible options are an increase in the training time, preprocessing of inputs, use of a different ANN paradigm, or use of different training parameters. Once the network has been trained, the mapping between the FLS inputs and the corresponding outputs is available in the form of the parameters of the ANN.

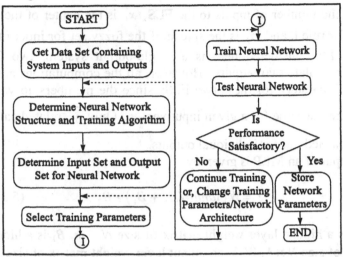

Figure 5.3: Design Procedure for an ANN Application.

The complexity of a handoff algorithm can be analyzed in terms of storage requirements and computational requirements. At every sampling instant, the outputs of the adaptive mechanisms (FLS, MLP, and RBFN) need to be calculated. See Chapter 2 for details on the FLS, MLP, and RBFN.

The output of an FLS, y, is given by

$$y = \frac{\sum_{l=1}^{R} \overline{y}^{l} \left(\mu_{B^l} \left(\overline{y}^{l} \right) \right)}{\sum_{l=1}^{R} \left(\mu_{B^l} \left(\overline{y}^{l} \right) \right)} \qquad (5.1)$$

where R is the number of rules, $\overline{y}^{l}$ is the center of the output fuzzy set, and $\mu_{B^l} \left(\overline{y}^{l} \right)$ (where B^l is a fuzzy set that is the output of the fuzzy inference engine) is calculated as

$$\mu_{B^l}\left(\overline{y}^l\right) = \prod_{i=1}^{m} exp\left(-\left(\frac{x_i - \overline{x}_i^l}{\sigma_i^l}\right)^2\right) \qquad (5.2)$$

where m is the number of inputs to the FLS, $\overline{x}_i^l$ is the center of the fuzzy set for input i for rule l, and σ_i^l is the spread of the fuzzy set for input i for rule l. Since the FLS considered here is a single-output system, two FLSs are required to calculate two outputs. However, all the computations need not be carried out separately for these two FLSs since the membership values, i.e., $\mu_{B^l}\left(\overline{y}^l\right)$, are the same for a given input vector. Only $\overline{y}^l$ related calculations need to be carried out for individual outputs.

The output of an MLP is given by

$$Y = W_2\, tanh(W_1 X + B_1) + B_2 \qquad (5.3)$$

where W_1 is a hidden layer weight matrix of size $N \times m$, B_1 is a hidden layer bias matrix of size $N \times 1$, W_2 is an output layer weight matrix of size $p \times N$, B_2 is an output layer bias matrix of size $p \times 1$, X is an $m \times 1$ input vector, Y is a $p \times 1$ output vector, and *tanh* is the hyperbolic tangent function.

The output of an RBFN is given by

$$Y = W_2 A_1 + B_2 \qquad (5.4)$$

with

$$A_1 = radbas(dist(W_1, X) \times B_1). \qquad (5.5)$$

Here, W_1 consists of centers of Gaussian functions and is of size $N \times m$, B_1 consists of spreads associated with Gaussian functions and is of size $N \times 1$, W_2 is an output layer weight matrix of size $p \times N$, B_2 is an output layer bias matrix of size $p \times 1$, X is an $m \times 1$ input vector, Y is a $p \times 1$ output vector, and *dist* is the distance between X and each row of W_1. In other words,

$$dist(w, X) = \sqrt{\sum_{j=1}^{p} (w_j - X_j)^2}. \qquad (5.6)$$

Here, w represents one row of W_1.

"Radbas" is the radial-basis function given by

$$radbas(x) = exp(-x^2). \qquad (5.7)$$

The storage requirements of the FLS, MLP, and RBFN are derived next.

- **FLS Storage Requirements.** For an FLS rule, m centers ($\overline{x}_i^l$) and m spreads (σ_i^l) for the antecedent part of the rule and p centers ($\overline{y}^l$) for the consequent part of the rule are required. Thus, for each rule, a total of $(2m + p)$ elements are required (Eq. 5.1 and 5.2). Since there are R rules in an FLS, a total of $(2m + p)R$ elements need to be stored for the FLS.

- **MLP Storage Requirements.** For an MLP, W_1, B_1, W_2, and B_2 are required. Since W_1 is of size $N \times m$ and B_1 is of size $N \times 1$, the number of elements for the first layer is $Nm + N$. Since W_2 is of size $p \times N$ and B_2 is of size $p \times 1$, the number of elements for the second layer is $pN + p$. The total number of elements are $Nm + N + pN + p = m(N + 1) + p(N + 1) = (m + p)(N + 1)$.

- **RBFN Storage Requirements.** For an RBFN, the parameters W_1, B_1, W_2, and B_2 have the same dimensions as in the case of an MLP. Hence, a total of $(m + p)(N + 1)$ elements are needed for RBFN.

Table 5.1 summarizes the storage requirements for an FLS, MLP, and RBFN.

Table 5.1: Storage Complexity of Adaptation Mechanisms.

System	Elements
FLS	$(2m + p)\,R$
MLP	$(m + p)\,(N + 1)$
RBFN	$(m + p)\,(N + 1)$

In this chapter, $R = 27$, $m = 3$, $p = 2$, $N = 5$ for BPNN, and $N = 8$ for RBFN. Table 5.2 gives the improvement in storage requirements for the neural techniques. The MLP and RBFN give an improvement of the factor 7.2 (i.e., 216/30) and 4.8 (i.e., 216/45) over the FLS for storage requirements.

Table 5.2: Specific Examples of Storage Complexity.

System	Elements
FLS	216
MLP ($N = 5$)	30
RBFN ($N = 8$)	45

The computational complexities of the FLS, MLP, and RBFN are derived next. The operations of addition and subtraction are grouped together and referred to as adds. Also, the operations of multiplication and division are grouped together and referred to as multiplies. Evaluations of functions (such as exponential and square-root) are referred to as functions.

- **FLS Computations.** For each rule, the input vector is processed by Eq. 5.2. There are one subtraction, one division, one multiplication (squaring operation), and one function (exponential) evaluation for each element of the input vector X that has m elements. Thus, there are one add, two multiplies, and one function. For each rule, there are m adds, $2m$ multiplies, and m functions due to m inputs. The m exponential terms and $\overline{y}^{\,l}$ are multiplied, requiring m more multiplies. Hence, there are mR adds, $3mR$ multiplies, and mR functions for R rules. There are $(R - 1)$ additions (of $\overline{y}^{\,l}$ and $\mu_{B^l}\left(\overline{y}^{\,l}\right)$) in the numerator and one division of the numerator and denominator in Eq. 5.1. Thus, there are $Rm + (R-1) = (m + 1)R - 1$ adds, $3mR + 1$ multiplies, and mR functions for one output calculation. For each additional output, there are R multiplies (of $\overline{y}^{\,l}$ and $\mu_{B^l}\left(\overline{y}^{\,l}\right)$) and $R-1$ adds (of the terms $\overline{y}^{\,l}\,\mu_{B^l}\left(\overline{y}^{\,l}\right)$ and one numerator-denominator division). Hence, there are additional $R + 1$ multiplies and $R - 1$ adds for each additional output. If there are p outputs, $(R + 1)(p - 1)$ additional multiplies and $(R - 1)(p - 1)$ additional adds are required. Thus, the total number of adds is $(m + 1)R - 1 + (R - 1)(p - 1) = mR + R - 1 + Rp - R - p + 1 = mR + Rp - p$, the total number of multiplies is $3mR + 1 + (R + 1)(p - 1) = 3mR + 1 + Rp - R + p - 1 = (3m + p - 1)R + p$, and the total number of functions is mR.

- **MLP Computations.** Each row of $W_1 X$ multiplication requires m multiplies and $m - 1$ adds. Since there are N rows in W_1, a total of mN multiplies and $(m - 1)N$ adds are required. $W_1 X$ and B_1 addition requires N more adds. Thus, $(m - 1)N + N = mN$ adds are required. Hence, mN multiplies, mN adds, and N functions (*tanh* calculations) are required to carry out $tanh(W_1 X + B_1)$ operation. The multiplication of W_2 and *tanh* terms is between the $p \times N$ and $N \times 1$ matrices, requiring pN multiplies and $(p-1)N$ adds. The result of this multiplication is added to the $p \times 1$ matrix B_2, requiring additional p adds. Hence, the total number of adds are $mN + pN = (m + p)N$, and the total number of multiplies are $mN + pN = (m + p)N$.

- **RBFN Computations.** There are m subtractions, m multiplies (squaring operation), $m - 1$ adds, and one function (square-root) for each row of W_1 in Eq. 5.6. Hence, there are $2m - 1$ adds, m multiplies, and one function for Eq. 5.6. For N rows of W_1, there are $(2m - 1)N$ adds, mN multiplies, and N functions. The distance function is $N \times 1$, and it is added to B_1 of size $N \times 1$. This requires N additions. Hence, there are $N(2m - 1) + N = 2mN$ adds, Nm multiplies, and N functions for the argument of *radbas* in Eq. 5.5. The $N \times 1$ matrix is processed by radial basis functions. In each radial function, there is one multiplication (squaring) and one function evaluation (exponential) operation. Thus, there are N more multiplies and N more function evaluations. Hence, there are $2mN$ adds, $Nm + N$ multiplies, and $2N$ functions for the calculation of A_1. The evaluation of Eq. 5.4 requires pN multiplies and pN adds. Hence, there are a total of $2mN + pN = N(2m + p)$ adds, $Nm + N + pN = (m + p + 1)N$ multiplies, and $2N$ functions.

Table 5.3 summarizes the computational requirements of the FLS, MLP, and RBFN. Table 5.4 shows the improvement in computational requirements for the neural methods. There is an improvement of 8.8 (486/55 = 8.8) and 3.8 (486/128 = 3.8) for MLP and RBFN, respectively, over the FLS.

Table 5.3: Computational Complexity of Adaptation Mechanisms.

System	Multiplies	Adds	Function Evaluations
FLS	$3m + R + p$	$M + 2R + (p - 1)R$	mR
MLP	$N(m + p)$	$N(m - 1) + p(N - 1) + N + p$	N
RBFN	$N(p + m + 1)$	$N(p + 2m)$	$2N$

Table 5.4: Computational Complexity of Adaptation Mechanisms.

System	Multiplies	Adds	Function Evaluations	Total Operations
FLS	272	133	81	486
MLP	25	25	5	55
RBFN	48	64	16	128

5.3   PERFORMANCE EVALUATION

Figure 5.4 shows the input data of the training data set. These different combinations of inputs are applied to the FLS, and the corresponding FLS outputs are calculated. The FLS outputs constitute the output data of the training data set. This output data serves as the target or desired output when a neural network is trained. The ranges of SIR, traffic difference, and velocity are from 15 dB to 21 dB, -2 to 2 calls, and 20 m/s to 38 m/s, respectively. Note that there is no relation between the data points corresponding to the consecutive indices. The continuous lines are used only to connect the (unrelated) data points. The lines do not convey any trends.

Figure 5.5 shows the input test data used for testing the generalization property of the trained ANNs. The test inputs were generated randomly within the specified ranges of the variables. The corresponding FLS outputs are the desired outputs. Hence, if an ANN has learned the input-output mapping of the FLS well, the presentation of the test inputs shown here produces the outputs that are similar to the desired test outputs.

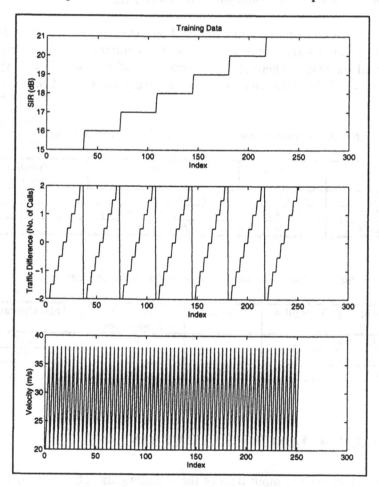

Figure 5.4: Training Data for Neural Networks.

Table 5.5 summarizes some of the results for MLP. E_{train} is the Frobenius norm of the difference between the desired outputs and the outputs of the MLP for the training data. E_{test} is the Frobenius norm of the difference between the desired outputs and the outputs of the MLP for the test data. A different number of hidden layer neurons were trained for different training times (5000 to 15000 epochs). One epoch is one pass through the training set. In general, more neurons can lead to an improved mapping. However, complexity increases as the number of neurons increases. The number of neurons is chosen to be eight as a compromise between the accuracy of generalization and complexity. Since the main interest is to represent the FLS with as few neurons as possible, a tradeoff between the number of neurons and the mapping accuracy must be achieved. For the application under consideration, the error performance of the MLP is quite acceptable.

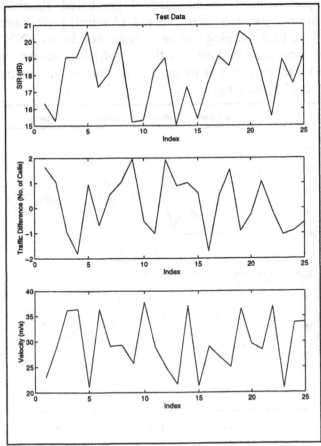

Figure 5.5: Test Data for Neural Networks.

Table 5.5: Training and Test Results for MLP.

Number	Number of Hidden Layer Neurons	E_{train}	E_{test}
1	5	24.58	8.17
2	8	16.32	8.33
3	12	16.01	8.09
4	17	16.47	8.07
5	21	16.22	8.06

Figure 5.6 shows the desired (or actual) test data and the MLP output data. The desired data and MLP predicted data are close to each other; however, they are not identical. This means that the ANN has learned most of the FLS mapping features, but it has not learned a perfect mapping. Hence, when the performances of the fuzzy logic based algorithm and neural algorithm are compared, similar, but not identical, performances should be expected. The FLS mapping can be learned well by an ANN provided that appropriate processing is done and that a sufficient number of neurons are used. Since the goal here is to use fewer neurons, no attempt is made to obtain perfect FLS mapping.

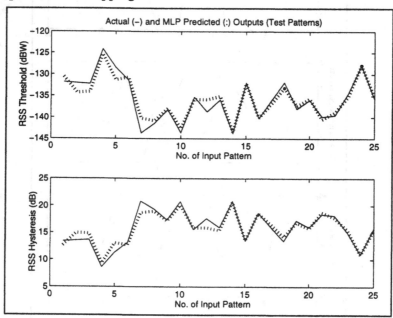

Figure 5.6: MLP Test Data Performance.

The RBFN mapping performance is similar to the MLP training performance. The RBFN has learned the mapping well, though the mapping is not exact. Table 5.6 summarizes some of the results for RBFN.

Table 5.6: Training and Test Results for RBFN.

Number	Number of Radial Basis Functions	E_{train}	E_{test}
1	5	34.50	8.28
2	10	24.42	6.82
3	14	20.25	5.47
4	19	17.80	5.29
5	24	15.73	5.16

Different spreads for radial basis functions and different numbers of radial basis functions were tried. In general, higher numbers of radial basis functions give an improved performance with the associated increase in complexity. The number of radial basis functions is chosen to be ten as a compromise between accuracy of generalization and complexity. As in the case of MLP, the performances of the fuzzy logic based algorithm and neural algorithm can be expected to be similar but not identical.

The CDF of RSS for conventional, fuzzy logic (FL), MLP, and RBFN algorithms is found to be almost identical. The FL, MLP, and RBFN algorithms provide a 0.8 dB improvement in the SIR distribution compared to the conventional algorithm. Again, there is a close match between the SIR distributions for FL, MLP, and RBFN algorithms. As an example of the performance similarity between the FL and the MLP algorithms, Figure 5.7 shows the traffic distribution. The FL and MLP algorithms give about a two-call improvement compared to the conventional algorithm.

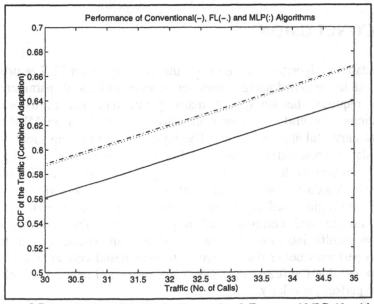

Figure 5.7: Distribution of Traffic for Conventional, Fuzzy, and MLP Algorithms.

Similarly, Figure 5.8 shows how close the performances of the FL and the RBFN algorithms are. Both the FL and RBFN algorithms provide a two-call improvement over the conventional algorithm. The operating points for the FL, MLP, and the RBFN algorithms are also found to be close to one another.

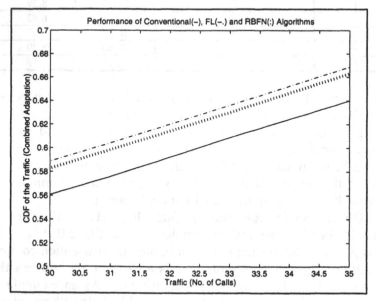

Figure 5.8: Distribution of Traffic for Conventional, Fuzzy, and RBFN Algorithms.

5.4   CONCLUSION

An adaptive algorithm that encodes the working of an FLS is proposed. An FLS can be designed using known sensitivities of handoff parameters, but the FLS requires the storage of many parameters and needs a lot of computations. Several neural network paradigms such as an MLP and an RBFN are universal approximators. The input-output mapping capability and compact data representation capability of these neural network paradigms are exploited to represent the FLS. The neural representation of the FLS provides an adaptive handoff algorithm that retains the high performance of the original fuzzy logic based algorithm and that has an efficient architecture for meeting storage and computational requirements. The analysis of the simulation results indicates that an adaptive multicriteria neural handoff algorithm performs better than a signal strength based conventional handoff algorithm and that the fuzzy logic based algorithm and neural network based algorithm perform similarly.

Chapter 6

A UNIFIED HANDOFF CANDIDACY ALGORITHM

This chapter proposes a new fuzzy logic based algorithm with a unified handoff candidate selection criterion and adaptive direction-biasing. The unified handoff candidate selection criterion allows the simultaneous consideration of several handoff criteria to select the best handoff candidate under given constraints. This chapter shows that adaptive direction-biasing improves the performance of the basic fuzzy handoff algorithm and allows additional degrees of freedom in achieving the desired balance among various system characteristics of interest.

6.1 AN ADAPTIVE FUZZY HANDOFF ALGORITHM WITH ADAPTIVE DIRECTION-BIASING

The algorithm proposed in this chapter incorporates adaptive direction-biasing into the fuzzy logic algorithm proposed in [117] and described in Chapter 4. Direction-biasing facilitates fast handoff. A direction-biased handoff algorithm for a microcellular environment is proposed in [60]. Since microcells frequently encounter the corner effect, fast moving vehicles must be connected to an umbrella cell or effective handoff algorithms must be used. A direction-biased handoff algorithm represents such an alternative solution [60] and has several nice features. Direction-biasing improves cell membership properties and handoff performance in LOS and NLOS scenarios in a multi-cell environment. The direction biased algorithm reduces the probability of dropped calls for hard handoffs, e.g., for TDMA systems, and reduces the time a user needs to be connected to more than one base station for soft handoffs, e.g., for CDMA systems, allowing more potential users per cell.

The basic idea behind the direction-biased algorithm is that handoffs to the BSs toward which the MS is moving are encouraged, while handoffs to the BSs from which the MS is receding are discouraged. Figure 6.1 shows the flowchart of a direction-biased algorithm.

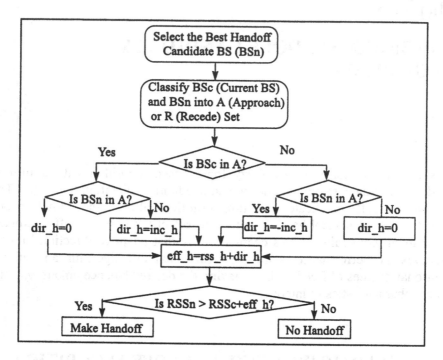

Figure 6.1: A Direction-Biased Algorithm.

First, the direction-biased algorithm selects the best handoff candidate BS based on the link measurements. For example, if a BS (BSn, where *n* is the BS identification) gives maximum RSS, it is selected as the best handoff candidate BS. The currently serving BS (BSc) and BSn are classified into one of the sets, set "Approach" (set A) or set "Recede" (set R). If the MS is moving toward a BS, this BS is classified into set A. If the MS is moving away from a BS, this BS is classified into set R. If both BSc and BSn are in set A or set R, the effective hysteresis (eff_h) is kept the same as the normal hysteresis value (rss_h). If BSc is in set A and BSn is in set R, the effective hysteresis is increased by the amount dir_h (which represents the amount of direction biasing). On the other hand, if BSc is in set R and BSn is in set A, the effective hysteresis is reduced by the amount dir_h.

A variation of the basic direction-biased algorithm is the preselection direction-biased algorithm [60]. If the best BS is a receding one and has a quality only slightly better than the second best BS, which is being approached, the handoff should be made to the second best BS because it is more likely to improve its chances of being selected in future. This rule provides a fast handoff algorithm with good cell membership properties without the undesirable effects associated with large hysteresis. Figure 6.2 shows the flowchart of the preprocessing for the preselection direction-biased algorithm. This algorithm processes the link measurements to bias the best handoff candidate selection process. A BS is classified into set A or set R based on the direction of the MS travel relative to the BS. If the BS is in set A, the link quality measurement is enhanced by a preselection hysteresis (Hp) to improve the chances of the selection of the BSs in set A as handoff candidates. However, if the BS is in set R, the link quality measurement is reduced by a preselection hysteresis (Hp) to deny the chances of the selection of the BSs in set R as handoff candidates. Once the link measurements of all the BSs are preprocessed, a BS with the best preselection-biased link measurement is selected as the best handoff candidate. Then, the normal direction-biased algorithm is executed.

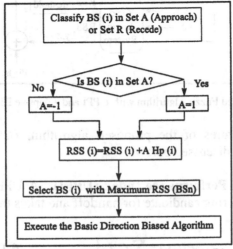

Figure 6.2: Preprocessing for Preselection Direction Biased Algorithm.

A pure direction-biased handoff algorithm has several disadvantages. This algorithm can lead to unnecessary handoffs (i.e., this algorithm may make a handoff even if the current BS provides good quality). The constant direction-biasing may cause this algorithm to make premature handoffs, potentially increasing the MS transmit power and causing high uplink interference.

Since direction-biasing has certain nice features, an adaptive version of direction-biasing has been incorporated into the basic fuzzy algorithm of [117]. A unified preselection performance index (UPPI) has been formulated to simultaneously consider several handoff criteria. Figure 6.3 shows the block diagram of the proposed fuzzy algorithm with UPPI and adaptive direction-biasing. First, the best handoff candidate is selected using the UPPI. Several measurements (such as RSS_c, RSS_n, and SIR_c) are averaged using the velocity adaptive averaging mechanism and processed by the direction-biased algorithm. This direction-biased algorithm uses two adaptive parameters

($RSS_{threshold}$ and $RSS_{hysteresis}$) supplied by an FLS. The value of RSS_{hyst} is modified based on adaptive direction biased hysteresis value (dir_h) to generate $RSS_{hysteresis}$. The mechanism of generating dir_h is explained later.

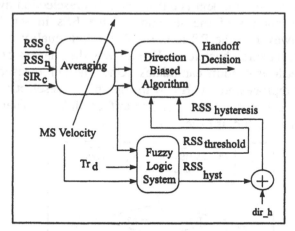

Figure 6.3: Proposed Fuzzy Algorithm with UPPI and Adaptive Direction-Biasing.

Two distinct features of the proposed algorithm, *UPPI* and *adaptive direction-biasing*, are discussed below.

Unified Preselection Performance Index (UPPI). A UPPI is proposed to select the best neighboring candidate for handoff and takes the form of

$$UPPI(j) = \sum_{i=1}^{N} C_i L_i^{2} \qquad (6.1)$$

where L_i is the ith *preprocessed* link measurement, N is the total number of link measurements (available from each potential handoff candidate BS), C_i is the weight given to the associated link measurement, and j is the BS index. A BS that gives the maximum UPPI is selected as the best handoff candidate. Preprocessing is necessary to normalize the measurements and to ensure that maximizing the link measurement related component of the UPPI indeed leads to the maximization of the UPPI. For example, it is desirable to minimize Tr_n (the number of ongoing calls in a cell) so that handoff blocking and call blocking probabilities are reduced. Since the goal is to maximize UPPI, Tr_n must be preprocessed so that the Tr_n related component of the UPPI increases when Tr_n decreases. The coefficients C_i ($i = 1, 2, ..., N$) can be adapted to reflect dynamics of the cellular environment. For example, when traffic intensity in a given service region is low, traffic balancing does not pay off. In such cases, it is desirable to reduce the weight of traffic in the overall

UPPI evaluation and avoid perturbing planned cell boundaries. On the other hand, when traffic intensity is high, traffic balancing reduces the handoff and call drop probability. The weight of traffic can be increased in such cases to achieve a higher degree of traffic balancing. Two link measurements of the neighboring cells, traffic in the cell and RSS from the BS, are used as L_i, and the weight of these L_i can be changed via coefficients (C_i). In other words,

$$UPPI(j) = C_1 L_1^2 + C_2 L_2^2 \qquad (6.2)$$

where L_1 is the preprocessed RSS for BS j and L_2 is the preprocessed traffic. When C_1 is zero, the selection of the best handoff candidate cell is based solely on traffic. When C_2 is zero, the selection of the best handoff candidate cell is based solely on RSS. Different weight can be given to RSS and traffic to obtain improved traffic balancing and adaptation. Since the basic fuzzy algorithm takes into account two neighboring cell measurements (RSS_n and Tr_n) for a handoff decision, this UPPI provides the best candidate in case handoff is necessary.

Adaptive Direction-Biasing Parameters. As explained earlier, the biasing influences handoff decisions even in the vicinity of the BS. However, a handoff should be discouraged even if the MS is receding from such a BS. Furthermore, the handoff region is located (approximately) midway between the BSs, and a higher degree of direction-biasing in the handoff region can prevent the ping-pong effect, reducing the number of handoffs. Based on these observations, direction-biasing parameters are adapted using Table 6.1. The input to the FLS is the difference between the distances of the MS from the serving BS and the candidate BS, and the output of the FLS is the incremental hysteresis value (dir_h). When the distance difference is high, i.e., when an MS is very close to one BS compared to another BS, there is a low degree of direction biasing. When the distance difference is low, i.e., when an MS is almost equidistant from both neighboring BSs, there is a high degree of direction-biasing.

Table 6.1: Fuzzy Rule Base for Adaptive Direction-Biasing.

Rule No.	Distance Difference	dir_h
1	High	Low
2	Normal	Normal
3	Low	High

6.2   PERFORMANCE ANALYSIS OF PROPOSED ALGORITHMS

The performances of the basic FL, direction-biased fuzzy logic (DBFL), and adaptive direction-biased fuzzy logic (ADBFL) algorithms are evaluated next.   A direction-biased algorithm that considers only RSS as a UPPI component is a DBFL algorithm, and a direction-biased algorithm that includes both RSS and traffic as UPPI components is a traffic direction-biased fuzzy logic (TDBFL) algorithm.   For FL and DBFL algorithms, C_1=1 and C_2=0.   For a TDBFL algorithm, C_1=0.7 and C_2=0.3.   Thus, in the case of the TDBFL algorithm, traffic is important in determining the best handoff candidate cell.

6.2.1   Performance Evaluation of FL, DBFL, and TDBFL Algorithms

The CDF of the RSS for FL, DBFL, and TDBFL algorithms is found to be similar.   There is a slight degradation in RSS distribution for the TDBFL algorithm (less than 0.1 dB).   This RSS degradation occurs because the selection of the handoff candidate cell is based on both the RSS and the traffic in the neighboring cells.   Hence, a slight degradation in RSS is expected.

Figure 6.4 shows the SIR distribution for FL, DBFL, and TDBFL algorithms. The FL and DBFL algorithms perform similarly, but there is a 0.5 dB degradation in SIR performance of TDBFL compared to FL and DBFL algorithms.   The slight degradation in RSS contributes to more degradation in SIR distribution for the TDBFL algorithm.

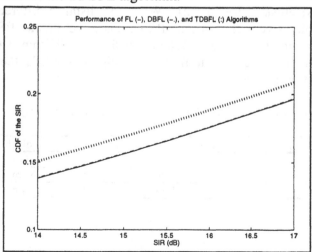

Figure 6.4: Distribution of SIR for FL, DBFL, and TDBFL Algorithms.

Figure 6.5 shows the traffic distribution for FL, DBFL, and TDBFL algorithms. Again, the performance of the FL and DBFL is similar, but there is a one call improvement for TDBFL compared to FL and DBFL algorithms. This result indicates that SIR performance can be traded for better traffic performance. This result has even more significant implications. Since the traffic performance is not critical under low traffic intensities, a UPPI that is receptive to RSS can be used to provide improved QoS quantified by voice quality and MS transmit power. When the traffic intensity is high, the UPPI should give more weight to traffic, and potentially, more users can be served by trading voice quality. Thus, an appropriate structure of the UPPI can help achieve high performance in the dynamic cellular environment.

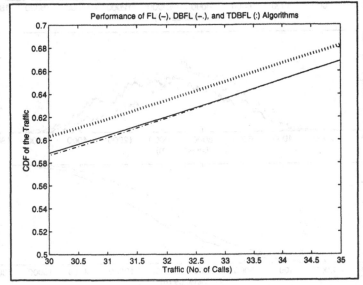

Figure 6.5: Distribution of Traffic for FL, DBFL, and TDBFL Algorithms.

Figure 6.6 shows the cell membership properties for FL, DBFL, and TDBFL algorithms. Pr(i) (i=0, 1, 2, 3) is the probability that the MS is connected to BS i. The MS travels from BS 0 to BS 2 at a constant velocity (65 mph). Pr(0) decreases from one to zero as the MS recedes from BS 0. Pr(1) increases from zero to one as the MS approaches BS 2. Both Pr(1) and Pr(3) increase until the midpoint of the MS's journey since the MS is moving toward these BSs. However, after the midpoint of the MS's journey, Pr(1) and Pr(3) decrease since the MS is now moving away from these BSs. As expected, the DBFL algorithm improves the cell membership properties of the basic FL algorithm. Corroborating this statement, Pr(0) and Pr(2) for the DBFL algorithm are lower and higher respectively than those for the FL algorithm. Moreover, Pr(1) and Pr(3) are higher than those for the FL algorithm until the midpoint of the MS's journey. Also, Pr(1) and Pr(3) are lower than those for the FL algorithm after the midpoint of the MS's journey. There is not much difference between the DBFL and TDBFL algorithm performance.

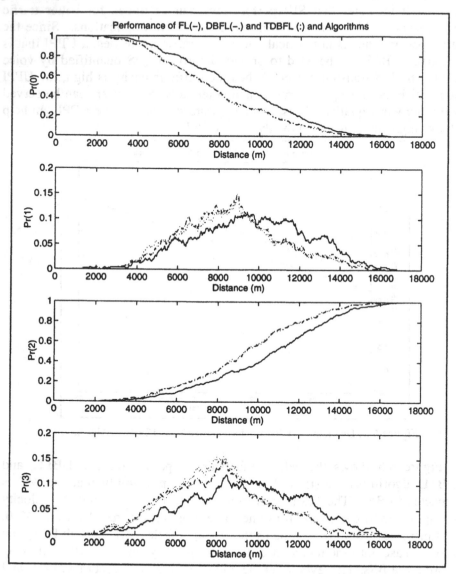

Figure 6.6: Cell Memberships for FL, DBFL, and TDBFL Algorithms.

Figure 6.7 shows the operating points for FL, DBFL, and TDBFL algorithms for different velocities. The DBFL algorithm gives fewer handoffs and less cross-over distance than the FL algorithm. The TDBFL gives even fewer handoffs than DBFL algorithm because traffic is important in the handoff candidate selection process for the TDBFL and because these traffic variations are less intense than RSS variations in the simulation environment. The mid-point of the MS's journey is 9 km. Recall that the operating point is defined by the average number of handoffs and the 50% cross-over distance.

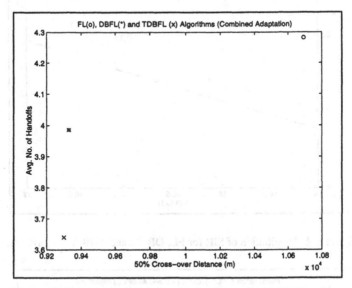

Figure 6.7: Operating Points for FL, DBFL, and TDBFL Algorithms.

6.2.2 Performance Evaluation of FL, DBFL, and ADBFL Algorithms

The RSS distribution is identical for FL, DBFL, and ADBFL algorithms. Figure 6.8 shows the SIR distribution for FL, DBFL, and ADBFL algorithms. There is a slight degradation (0.1 dB) in SIR performance of ADBFL compared to FL and DBFL algorithms since, in the handoff region, a cell with slightly better RSS may be available but will not be selected due to increased direction-biasing. Since the aim is to reduce the number of handoffs due to fading in the handoff region, RSS and SIR are traded off to obtain fewer handoffs.

Figure 6.9 shows the traffic distribution for FL, DBFL, and ADBFL algorithms. The distribution is identical for DBFL and ADBFL algorithms and shows an improvement over the FL algorithm.

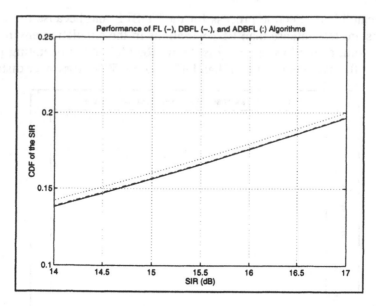

Figure 6.8: Distribution of SIR for FL, DBFL, and ADBFL Algorithms.

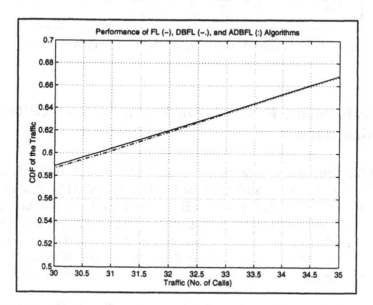

Figure 6.9: Distribution of Traffic for FL, DBFL, and ADBFL Algorithms.

Figure 6.10 shows cell memberships for FL, DBFL, and ADBFL algorithms. This figure shows the distinct advantage of ADBFL over FL and DBFL algorithms; ADBFL improves cell membership properties. This improvement is reflected in fewer handoffs and reduced cross-over distance.

Pr(0), Pr(2), and Pr(3) decrease quickly and Pr(4) increases quickly after the midpoint of the MS journey.

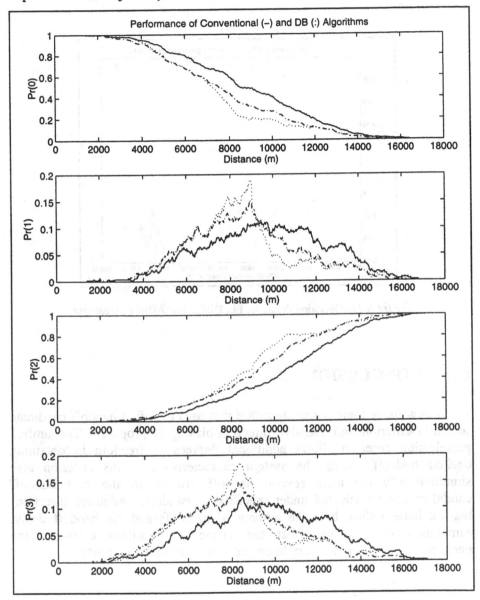

Figure 6.10: Cell Memberships for FL, DBFL, and ADBFL Algorithms.

Figure 6.11 shows the operating points for FL, DBFL, and ADBFL algorithms. DBFL and ADBFL provide much better performance than the FL algorithm. Also, the ADBFL algorithm gives fewer handoffs and reduced crossover distance compared to the DBFL algorithm. Thus, the ADBFL algorithm can help achieve an optimum operating point in a given environment.

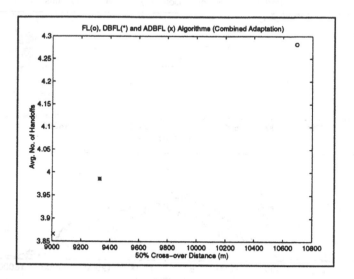

Figure 6.11: Operating Points for FL, DBFL, and ADBFL Algorithms.

6.3   CONCLUSION

A new fuzzy logic based algorithm that uses a unified handoff candidate selection criterion and adaptive direction-biasing is proposed. The unified preselection criterion allows additional degrees of freedom in obtaining desired tradeoff among the system characteristics. This criterion also simultaneously considers several handoff criteria so the best handoff candidate can be selected under specified constraints. Adaptive direction-biasing helps reduce both the number of handoffs and the handoff delay. Simulation results show that the proposed algorithm enhances the performance of the basic fuzzy logic and direction-biased algorithms.

Chapter 7

PATTERN CLASSIFICATION BASED HANDOFF ALGORITHMS

A new class of adaptive handoff algorithms that views the handoff problem as a pattern classification problem is proposed in this chapter. Neural networks and fuzzy logic systems are good candidates for pattern classifiers due to their nonlinearity and generalization capability. Simulation results show that the proposed algorithms improve the distributions of SIR and traffic compared to the conventional algorithm, increasing the spectral efficiency and quality of service of the cellular system. Adaptive direction-biasing is proposed to reduce the processing load and improve the cell membership properties.

7.1 HANDOFF AS A PATTERN CLASSIFICATION PROBLEM

A multiple criteria handoff algorithm can provide better performance than a single criterion handoff algorithm due to additional degrees of freedom and to a greater potential for achieving the desired balance among different system characteristics. Pattern classification (or pattern recognition (PR)) is a convenient and compact way of implementing a multicriteria handoff algorithm. The PR based handoff algorithms are introduced in Chapter 1 (see Section 1.6.2).

Fuzzy logic is a good candidate as a pattern classifier (PC) for several reasons. An FLS can act as a universal approximator and, hence, can mimic the working of an ideal PC by learning the relationships among the variables of a training data set. There is an inherent fuzziness in the actual cell boundaries due to the dynamics of the cellular environment, and, by nature, an FLS can model this fuzziness. The concept of the degree of membership in fuzzy logic is very similar to the PC concept of the degree to which a pattern belongs to a class.

Neural networks are also good candidates as a PC. Several paradigms of neural networks can act as a universal approximator and, hence, can mimic the working of an ideal PC through supervised learning on the training data set. The dynamics of the cellular environment are very complex, and the nonlinearity of neural networks can model such enormous complexity. The PR concept of the degree to which a pattern belongs to a class can be learned by a neural network. Figure 7.1 shows the block diagram of a pattern classifier.

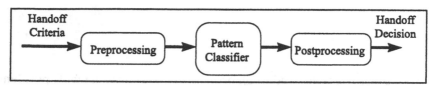

Figure 7.1: Pattern Classification Based Handoff Algorithm.

A set of handoff criteria is processed to create a pattern vector. This vector is classified into one of the classes. The class may represent the BS identification, or it may represent the degree to which the MS belongs to a particular BS. The output of this PC can be postprocessed to decide if handoff is required. The PR approach to handoff has several advantages. The multicriteria nature of PR allows for the simultaneous consideration of several significant aspects of the handoff procedure to enhance the system performance in accordance with the defined global system goals. PR is a direct approach for handoff in which testing of a sequence of binary IF-THEN rules of a conventional algorithm is replaced by a single operation of classification. A PC has a high potential for parallel implementation, which facilitates implementation of a fast handoff algorithm. A PC is inherently a single output system, and hence it is relatively less complex than the multiple output mechanisms used as part of adaptive multicriteria handoff algorithms, which can lead to an improvement in both computational and storage requirements. Adaptation capability can be easily built into the PC by appropriately designing the PC (e.g., by choosing appropriate decision rules for the PC). Useful features of existing handoff algorithms can be easily incorporated into the PC design by properly preprocessing the handoff criteria.

Clustering algorithms and *PR algorithms* are applied to RSS measurements for determining the service area of BSs in [64]. Clustering algorithms utilize clusters, geometrical regions where data points are concentrated according to distance measures, to assign membership values to the input pattern. The clustering algorithm works on a finite data set, and clusters evolve based on this data set. A PR algorithm assigns the input pattern to a class based on the explicit or implicit decision rules that define boundaries between the classification regions. The PC is based on supervised learning, and it can assign membership values to new patterns after it is trained. The PR algorithm proposed in [64] requires the knowledge of the measurement statistics. A PR algorithm that overcomes this drawback and exploits several features of fuzzy logic and neural networks to provide a high performance handoff algorithm is proposed in this chapter. The proposed algorithm exploits the strengths of a full-fledged FLS rather than relying only on the concept of the degree of membership.

A generic procedure for implementing an adaptive multicriteria handoff algorithm in a PC framework is described in Section 7.2. The performances of a conventional algorithm and the proposed neural algorithm are evaluated in Section 7.3. Finally, Section 7.4 summarizes the chapter.

7.2 DESIGN OF A PATTERN CLASSIFIER FOR HANDOFF

Figure 7.2 shows three distinct phases involved in the design of a PC. The phases include determination of the training data set, determination of a PC structure, and actual operation of classification. These phases and the specific PR based handoff algorithms are discussed next.

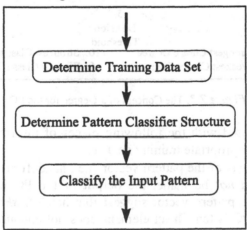

Figure 7.2: Phases of Pattern Classifier Design.

7.2.1 Determination of the Training Data Set

The first step is to create a training data set that consists of representative patterns and the corresponding class association degrees (i.e., the degree to which a given pattern belongs to a class). The following handoff criteria are used to form a pattern vector: RSS_c - $RSS_{threshold}$ (or ip_1), RSS_n - (RSS_c + $RSS_{hysteresis}$) (or ip_2), SIR_c - $SIR_{threshold}$ (or ip_3), $Tr_d = Tr_c$-Tr_n (or ip_4), and *MS velocity* (or ip_5). RSS_c is the received signal strength (RSS) from the current BS, $RSS_{threshold}$ is the RSS threshold, RSS_n is the received signal strength (RSS) from the neighboring (or candidate) BS, $RSS_{hysteresis}$ is the RSS hysteresis, SIR_c is the SIR of the current channel, $SIR_{threshold}$ is the SIR threshold, Tr_d is the traffic difference (i.e., the difference between the number of calls in the current cell (Tr_c) and the number of calls in the neighboring cell (Tr_n)).

Figure 7.3 illustrates the concept of the association degree for the PC. Class c denotes the class of the current BS, and class n denotes the class of the neighboring BS.

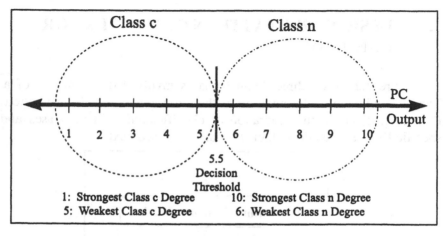

Figure 7.3: The Concept of a Degree for the PC.

Figure 7.3 is used with the following pieces of information to illustrate the creation of an appropriate training data set.

- If all the elements of the pattern vector (i.e., handoff criteria) suggest that handoff *should not* be made, the output of the PC is one. If all the elements of the pattern vector suggest that handoff *should* be made, the output of the PC is ten. If an element does not encourage or discourage handoff, its position is considered neutral.
- If the majority of the pattern elements favor a "No Handoff" decision, the output of the PC is in the range of one to five. Similarly, if the majority of the pattern elements favor a "Handoff" decision, the output of the PC is in the range of six to ten. The output value "1" indicates that the degree to which an MS belongs to class c is strongest, and the output value "5" indicates that the degree to which an MS belongs to class c is weakest. The output value "10" indicates that the degree to which an MS belongs to class n is strongest, and the output value "6" indicates that the degree to which an MS belongs to class n is weakest.

- The output of the PC depends on the net agreements between the elements of the pattern for a particular decision, "Handoff" or "No Handoff."
- The training data set should cover the entire range of interest for all the variables. To minimize the number of patterns in the training data set, representative examples should be chosen carefully.

Examples for handoff decisions are given below.

- **Examples for a "No Handoff" Decision.** If there are three agreements for "No Handoff" and two neutral positions, the net number of agreements is three and the PC output is three. If there are four neutral positions and one agreement, the net agreement is one and the output value is five. If there are four agreements and one disagreement, the net number of agreements is three and the output is three. Table 7.1 summarizes the PC outputs for the "No Handoff" decision scenario.
- **Examples for a "Handoff" Decision.** If there are three agreements for "Handoff" and two neutral positions, the net number of agreements is three and the output is eight. If there are four neutral positions and one agreement, the net agreement is one and the output value is six. If there are four agreements and one disagreement, the net number of agreements is three and the output is eight. Table 7.2 summarizes the PC outputs for the "Handoff" decision scenario.

Table 7.1: PC Outputs for "No Handoff" Decision.

Number	Net Agreements	PC Output
1	5	1
2	4	2
3	3	3
4	2	4
5	1	5

Table 7.2: PC Outputs for "Handoff" Decision.

Number	Net Agreements	PC Output
1	5	10
2	4	9
3	3	8
4	2	7
5	1	6

Several concepts of fuzzy logic were used to create a training data set. Each fuzzy variable (each element of the input pattern vector called ip_j, $j \in$ [1,5]) is divided into three fuzzy sets ("High" (H), "Medium" (M), and "Low"

(L)). Two basic rules are used to derive a complete set of the rules, covering the entire region of interest:

- **Rule 1.** If ip_1 is H, ip_2 is L, ip_3 is H, ip_4 is L, and ip_5 is L, the output is one;
- **Rule 2.** If ip_1 is L, ip_2 is H, ip_3 is L, ip_4 is H, and ip_5 is H, the output is ten.

Since there are five fuzzy variables and three fuzzy sets, there are a total of $3^5 = 243$ rules.

7.2.2 Pattern Classifier: Structure and Actual Operation of Classification

The input to the PC is a pattern vector, and the output of the PC is a value that indicates the degree to which the input pattern belongs to the class c (i.e., the current BS) and class n (i.e., the neighboring BS). A technique can be used to learn the training data relationships. Two architectures of neural networks, MLP and RBFN, and the Mamdani FLS are used as PC structures. The details of neural network paradigms can be found in Chapter 2 and [91], and the details of the Mamdani FLS can be found in Chapter 2 and [89].

The PC classifies the input vector into the class associated with the closest stored pattern vector. The closeness can be quantified by the Euclidean distance between the stored patterns and the input pattern vector. The PC output indicates the degree to which a given pattern vector belongs to a class (or a BS).

7.2.3 Details of the PC-Based Handoff Algorithms

Figure 7.4 shows the block diagram of a fuzzy logic PC-based (FLPC-based) handoff algorithm. The link measurements, RSS_c, RSS_n, and SIR_c, are averaged using a velocity adaptive averaging mechanism [60]. *MS velocity* and traffic difference Tr_d are not averaged since their instantaneous values are of interest. The averaged RSS_c, RSS_n, and SIR_c are biased before forming a pattern to account for the thresholds. The FLPC assigns a class association degree to the input pattern. If this degree is greater than 5.5, handoff is made.

Figure 7.5 shows the block diagram of a neural network PC-based handoff algorithm. This algorithm is similar to the FLPC algorithm except a neural network is used as a PC instead of an FLS.

Figure 7.6 shows how adaptive direction biasing is incorporated into the basic PC-based handoff algorithm. An adaptive direction biasing mechanism provides adaptive $RSS_{hysteresis}$, which is used to form an input pattern for the PC. In other respects, this algorithm is similar to the FLPC algorithm. Either an FLS or a neural network can be used as a PC.

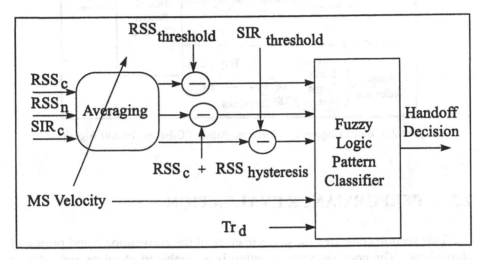

Figure 7.4: Block Diagram of an FLPC-Based Handoff Algorithm.

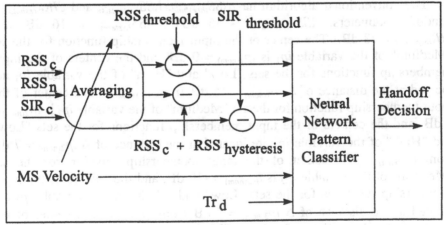

Figure 7.5: Block Diagram of a Neural Network PC-Based Handoff Algorithm.

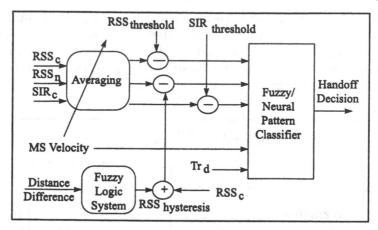

Figure 7.6: Block Diagram of a Direction-Biased PC-Based Handoff Algorithm.

7.3   PERFORMANCE EVALUATION

This section compares the performances of the conventional and proposed algorithms. The conventional algorithm is a combined absolute and relative signal strength based algorithm (see Chapter 4). The macrocellular simulation model described in Chapter 3 is used here to analyze the algorithms.

The conventional algorithm has $RSS_{threshold}$, $RSS_{hysteresis}$, and $SIR_{threshold}$ as handoff parameters. $RSS_{threshold}$ = -136 dBW, $RSS_{hysteresis}$ = 16 dB, and $SIR_{threshold}$ = 28 dB. The center of the input membership function for the set "Medium" of the variable ip_1 is $ip_{1(nom)}$ = 0 dB, and the centers of the input membership functions for the sets "Low" and "High" of the variable ip_1 are located at the distance of $\Delta ip_{1(nom)}$ = 15 dB from $ip_{1(nom)}$. The center of the input membership function for the set "Medium" of the variable ip_2 is $ip_{2(nom)}$ = 0 dB, and the centers of the input membership functions for the sets "Low" and "High" of the variable ip_2 are located at the distance of $\Delta ip_{2(nom)}$ = 7 dB from $ip_{2(nom)}$. The center of the input membership function for the set "Medium" of the variable ip_3 is $ip_{3(nom)}$ = -10 dB, and the centers of the input membership functions for the sets "Low" and "High" of the variable ip_3 are located at the distance of $\Delta ip_{3(nom)}$ = 5 dB from $ip_{3(nom)}$. The center of the input membership function for the set "Medium" of the variable ip_4 is $ip_{4(nom)}$ = 0, and the centers of the input membership functions for the sets "Low" and "High" of the variable ip_4 are located at the distance of $\Delta ip_{4(nom)}$ = 2 from $ip_{4(nom)}$. The center of the input membership function for the set "Medium" of the variable ip_5 is $ip_{5(nom)}$ = 29 m/sec, and the centers of the input membership functions for the sets "Low" and "High" of the variable ip_5 are located at the distance of $\Delta\ ip_{5(nom)} = 9$ m/sec from $ip_{5(nom)}$. The maximum output is ten, and the minimum output is one. Recall from Chapter 4 that the system should use "High" MS velocity in practice when the MS is moving *toward* the candidate BS at a high velocity. The spreads of the membership functions are chosen in such a way that the membership value drops to zero at the center of the membership function of the nearest set.

7.3.1 Evaluation of a Fuzzy Logic Pattern Classifier Handoff Algorithm

The CDF of RSS for the conventional and FLPC-based algorithms is similar and hence not shown here. Figure 7.7 shows the distribution of SIR for both the algorithms. The SIR distribution for the FLPC algorithm is improved by 1.3 dB compared to the conventional algorithm. This SIR improvement leads to better voice quality, fewer dropped calls, lower transmit power, and lower overall global interference level. This improvement in SIR is due to the interference adaptation of the proposed algorithm.

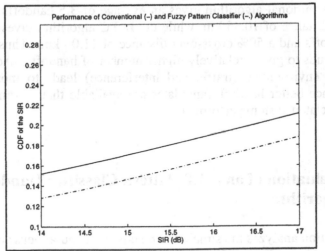

Figure 7.7: Distribution of SIR for Conventional and FLPC Algorithms.

Figure 7.8 shows the traffic distribution for the algorithms. The FLPC algorithm gives a 4.3 call improvement in the traffic distribution. In other words, the FLPC can accommodate 4.3 more users than the conventional algorithm, reducing call and handoff blocking probabilities and enhancing spectral efficiency of the cellular system. Traffic adaptation of the proposed algorithm provides this improvement in traffic distribution.

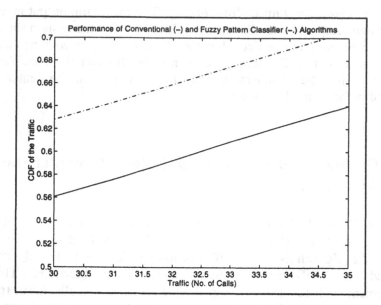

Figure 7.8: Distribution of Traffic for Conventional and FLPC Algorithms.

The conventional algorithm gives an average of 3.5 handoffs and a 50% cross-over distance of 10.85 km, while the FLPC algorithm gives an average of 6.2 handoffs and a 50% cross-over distance of 11.03 km. Thus, the FLPC algorithm tends to give a relatively higher number of handoffs. Adaptation to the cellular environment (traffic and interference) leads to more frequent handoffs since better handoff candidates are available that can improve SIR and traffic related system performance.

7.3.2 Evaluation of an MLP Pattern Classifier Handoff Algorithm

This section analyzes the simulation results for a neural network based PC when an MLP is used as a PC.

The CDF of RSS for the conventional and MLP PC algorithms are similar and not shown here. Figure 7.9 shows the distribution of SIR for both the algorithms. The MLP PC algorithm gives a 1.9 dB improvement in SIR over the conventional algorithm, improving QoS related system performance.

Figure 7.10 shows the traffic distribution for different algorithms. The MLP PC algorithm gives a 3.5 call better traffic distribution than the conventional algorithm, increasing the potential number of new users.

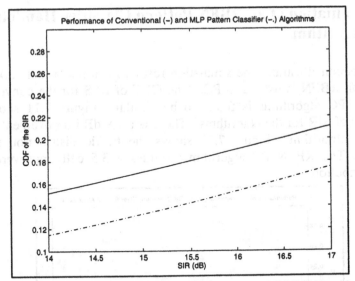

Figure 7.9: Distribution of SIR for Conventional and MLP PC Algorithms.

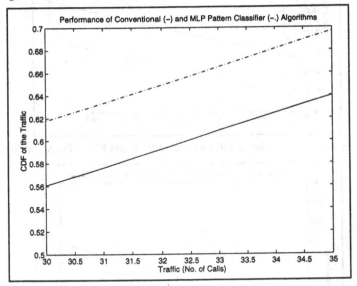

Figure 7.10: Distribution of Traffic for Conventional and MLP PC Algorithms.

The conventional algorithm gives an average of 3.5 handoffs and a 50% cross-over distance of 10.85 km, while the MLP PC algorithm gives an average of 7.3 handoffs and a 50% cross-over distance of 10.80 km. Thus, the MLP PC increases the network load but reduces interference.

7.3.3 Evaluation of an RBFN Pattern Classifier Handoff Algorithm

This section illustrates the simulation results for a neural network based PC when an RBFN is used as a PC. The CDF of RSS for the conventional and RBFN PC algorithms is found to be similar. Figure 7.11 shows the distribution of SIR for the algorithms. There is a 1.9 dB improvement for the RBFN PC algorithm. Figure 7.12 shows the traffic distribution for the algorithms. The RBFN PC algorithm provides a 3.5 call improvement in traffic distribution.

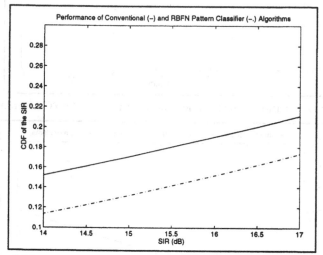

Figure 7.11: Distribution of SIR for Conventional and RBFN PC Algorithms.

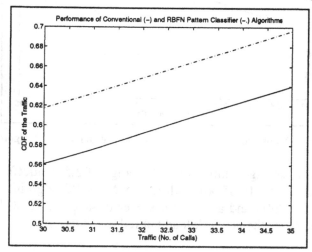

Figure 7.12: Distribution of Traffic for Conventional and RBFN PC Algorithms.

The conventional algorithm gives an average of 3.5 handoffs and a 50% cross-over distance of 10.85 km, while the RBFNPC algorithm gives an average of 7.5 handoffs and a 50% cross-over distance of 11.00 km.

7.3.4 Evaluation of a Direction-Biased MLP Pattern Classifier Handoff Algorithm

Similar CDF of RSS for the conventional and direction biased MLP PC algorithms is obtained. Figure 7.13 shows the distribution of SIR for the algorithms. The direction-biased MLP PC is 1.5 dB better than the conventional algorithm. Note the reduction in SIR improvement for the direction-biased MLP PC compared to the non-direction-biased MLP PC algorithm. As discussed earlier, adaptive direction-biasing can tradeoff SIR performance for a reduced number of handoffs and improved cell membership properties.

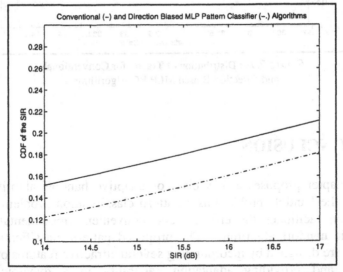

Figure 7.13: Distribution of SIR for Conventional and Direction-Biased MLP PC Algorithms.

Figure 7.14 shows the traffic distribution for two different algorithms. The direction-biased MLP PC algorithm gives a 3.5 call improvement in the traffic performance.

The conventional algorithm gives an average of 3.5 handoffs and a 50% cross-over distance of 10.85 km, the non-direction-biased MLP PC algorithm gives an average of 7.3 handoffs and a 50% cross-over distance of 10.80 km, and the direction-biased MLP PC algorithm gives an average of 6.5 handoffs and a 50% cross-over distance of 9.40 km. The distance of 9 km is midway between the BSs.

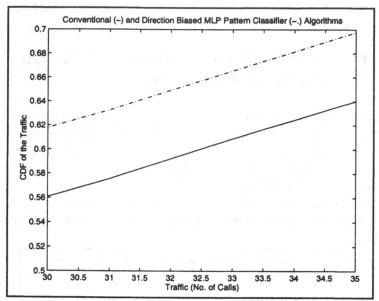

Figure 7.14: Distribution of Traffic for Conventional
and Direction-Biased MLP PC Algorithms.

7.4   CONCLUSION

This chapter proposes a new class of adaptive handoff algorithms that formulates the handoff problem as a pattern classification problem. Pattern classification facilitates the efficient and convenient implementation of a multicriteria handoff algorithm. The proposed pattern classification based algorithms are designed by incorporating several attractive features of existing algorithms and providing adaptation capability in a dynamic cellular environment through neural networks and fuzzy logic systems. Extensive simulation results for a conventional handoff algorithm (absolute and relative signal strength based algorithm) and pattern classification based algorithms are presented. Adaptive direction biasing is incorporated into the pattern classifier algorithms to reduce the processing load and improve the cell membership properties.

Chapter 8

MICROCELLULAR HANDOFF ALGORITHMS

Microcells increase system capacity but make resource management difficult. They impose distinct constraints on handoff algorithms due to the characteristics of the propagation environment. Adaptive handoff algorithms for microcells are proposed. The proposed non-direction-biased algorithm utilizes adaptive parameters supplied by a fuzzy logic system. Direction-biasing has been incorporated into the basic non-direction-biased algorithm to obtain a good basic adaptive algorithm suitable for a microcellular environment. Adaptation to traffic, interference, and mobility has been superimposed on the basic direction-biased algorithm. It is shown that the proposed algorithms provide high performance in generic handoff scenarios in a microcellular system.

8.1 INTRODUCTION TO HANDOFFS IN MICROCELLS

Microcells increase system capacity at the cost of an increase in the complexity of resource management. In particular, the number of handoffs per call increases, and fast handoff algorithms are required to maintain an acceptable level of dropped call rate. Microcells impose distinct constraints on handoff algorithms due to the characteristics of their propagation environment. For example, an MS encounters a propagation phenomenon called *corner effect*, which demands a faster handoff (see Chapter 1). Figure 8.1 shows two generic handoff scenarios in microcells, a *line-of-sight handoff* and a *non-line-of-sight handoff*. An LOS handoff occurs when the BSs that serve an MS are LOS BSs before and after the handoff. When the MS travels from BS 0 to BS 2, it experiences an LOS handoff. An NLOS handoff occurs when one BS is a NLOS BS before the handoff, and the other BS becomes an NLOS BS after the handoff. When the MS travels from BS 0 to BS 1, it experiences an NLOS handoff. A good handoff algorithm performs uniformly well in both generic handoff scenarios. Important considerations for designing handoff algorithms for a microcellular system are introduced next.

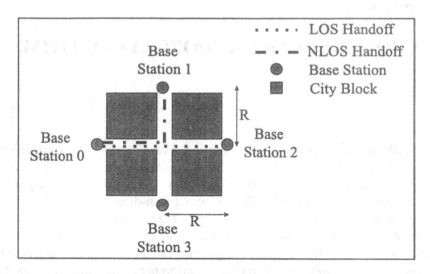

Figure 8.1: Generic Handoff Scenarios in a Microcellular System.

- **Mobility and Traffic Characteristics**. The MS speeds are lower, and the speed range is narrower compared to a macrocellular scenario. Traffic is normally allowed only along the streets.
- **Propagation Features**. The MS experiences the corner effect as discussed earlier. Field measurements have shown that the shadow fading intensity is lower in microcells than in macrocells.
- **Measurement Averaging**. The averaging interval (or averaging distance) should be shorter in microcells to respond to fast varying signal strength profiles. To provide adequate averaging to counteract shadow fading effects, a sufficient number of samples are required, which may necessitate higher measurement sampling frequency.
- **Primary Handoff Requirements**. A handoff algorithm should be fast and should minimize the number of handoffs. A fixed parameter handoff algorithm is suboptimal in a microcellular environment. For example, if hysteresis is large, it will cause a delay in NLOS handoff, increasing the probability of a dropped call. On the other hand, if hysteresis is small, it will increase the likelihood of the ping-pong effect. Since the situation of LOS or NLOS handoff cannot be known *a priori*, a proper tradeoff must be achieved between the LOS and NLOS handoff performance. In general, a large hysteresis gives good LOS handoff performance but poor NLOS handoff performance. A small hysteresis gives good NLOS handoff performance but poor LOS handoff performance.

- **Secondary Handoff Requirements**. The algorithm should respond relatively faster to fast moving vehicles, should attempt to balance traffic, and should be adaptive to interference.

The requirement of a fast handoff for a NLOS handoff case can be met by deploying a macrocellular overlay system over an existing microcellular system. However, this is an expensive solution, and it complicates even further the resource management of the already complex microcellular system. A better solution is to design a good handoff algorithm that can perform well in both LOS and NLOS handoff cases. Some research efforts have been made to cope with the problems associated with microcellular handoffs. Grimlund [19] evaluates an RSS-based algorithm with hysteresis for LOS and NLOS handoffs and indicates that fast handoffs can be made in a NLOS case at the cost of a higher number of handoffs in a LOS case. A handoff algorithm that consists of an "OR" circuit between two separate decision-making mechanisms for LOS and NLOS handoff cases is proposed in [118]. The LOS handoff decision-making mechanism uses longer averaging time and smaller hysteresis, while the NLOS handoff mechanism uses shorter averaging time and large hysteresis. The performance of this algorithm is velocity sensitive; the best handoff performance is obtained only at one velocity. Austin [58] overcomes this drawback by proposing velocity-adaptive handoff algorithms. Austin [60] develops direction biased handoff algorithms to improve handoff performance in LOS and NLOS cases. These algorithms are evaluated in a multi-cell environment (a four BS neighborhood model).

New adaptive handoff algorithms that perform well in both LOS and NLOS handoff situations are proposed. These new algorithms are described in Section 8.2, and Section 8.3 analyzes the performance of the algorithms from different significant aspects. Finally, Section 8.4 summarizes the chapter.

8.2 ADAPTIVE HANDOFF ALGORITHMS

This section describes the development of a generic microcellular algorithm that attempts to achieve a balance between LOS and NLOS handoff performance. The objective of this generic algorithm is to provide a good basic algorithm that meets the primary handoff requirements discussed earlier. Also described is an adaptive algorithm that attempts to meet secondary handoff requirements after making sure that the primary handoff requirements are satisfied.

8.2.1   A Generic Microcellular Algorithm

This section describes the development of a generic microcellular algorithm in two stages. First, a non-direction-biased algorithm is described, and then, direction-biasing is incorporated into the algorithm to provide a basic adaptive algorithm that can perform well in a typical microcellular environment.

Figure 8.2 shows the block diagram of an adaptive handoff algorithm suitable for a microcellular system.

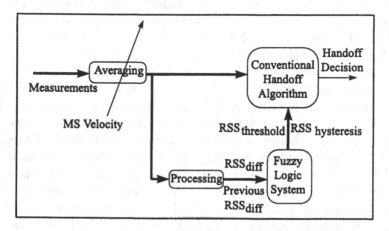

Figure 8.2: Block Diagram of an Adaptive Microcellular Handoff Algorithm.

Handoff criteria are averaged according to the velocity-adaptive averaging mechanism. The conventional algorithm is a combination of an absolute and relative RSS-based algorithm and an SIR-based algorithm. The RSS-based algorithm has threshold ($RSS_{threshold}$) and hysteresis ($RSS_{hysteresis}$) as parameters while the SIR based algorithm has threshold ($SIR_{threshold}$) as a parameter. If RSS drops below $RSS_{threshold}$ or SIR drops below $SIR_{threshold}$, a handoff process is initiated. If another BS can provide RSS that exceeds the RSS of the current BS by an amount $RSS_{hysteresis}$, a handoff is made to the new BS. The SIR threshold parameter allows the early initiation of a better handoff candidate search. The RSS-based parameters are adapted using an FLS. The input to the FLS is the difference in RSS between two best BSs (i.e., BSs from which the MS receives maximum RSSs). Generally, these BSs are LOS BSs since an MS has lower RSS from NLOS BSs. The outputs of the FLS are adaptive handoff parameters, $RSS_{threshold}$, and $RSS_{hysteresis}$. The geometry that underlines the philosophy behind the design of the FLS is illustrated in Figure 8.3.

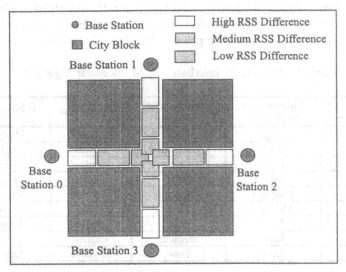

Figure 8.3: Handoff Situations in a Microcellular System.

When an MS is relatively far from the intersection and close to a BS, the difference in RSS at the MS from the LOS BSs is high since the MS receives very high RSS from the closer BS and very low RSS from the far LOS BS. This situation is similar to a LOS handoff situation since the good handoff candidates are LOS BSs. Under these circumstances, it is advantageous to use high $RSS_{hysteresis}$ and low $RSS_{threshold}$ values to reduce the ping-pong effect. However, as an MS reaches the intersection, there is a likelihood of a NLOS handoff, and it is beneficial to use low $RSS_{hysteresis}$ and high $RSS_{threshold}$ to make a fast handoff in case a NLOS handoff is necessary. The intersection region is characterized by small (ideally, near zero) RSS differences. After an MS crosses an intersection, the RSS difference keeps increasing, and this situation is similar to a LOS handoff scenario. Again, it is important to use high $RSS_{hysteresis}$ and low $RSS_{threshold}$ to reduce the ping-pong effect. Based on the knowledge of such propagation characteristics of a microcellular environment, a fuzzy logic rule base is created as shown in Table 8.1. *Current RSS Difference* and *Previous RSS Difference* are inputs to the rule base, and $RSS_{threshold}$ and $RSS_{hysteresis}$ are the outputs of the rule base. *Current RSS Difference* is the difference in RSS from two best BS at the current sample time, and *Previous RSS Difference* is the difference in RSS from two best BS at the previous sample time. Consider Rule 1. When *Current RSS Difference* and *Previous RSS Difference* are high, the MS is close to a BS, and, hence, $RSS_{threshold}$ is made lowest and $RSS_{hysteresis}$ is made highest to prevent the ping-pong effect. On the other hand, when *Current RSS Difference* and *Previous RSS Difference* are low, the MS is equally far from the BSs, and, hence, $RSS_{threshold}$ is made highest and $RSS_{hysteresis}$ is made lowest to make a fast handoff in potential NLOS handoff situations. The idea of using high hysteresis for LOS handoff situations and low hysteresis for NLOS handoff situations conforms to the primary handoff objectives.

Table 8.1: Fuzzy Logic Rule Base for a Microcellular Algorithm.

Rule Number	Current RSS Difference	Previous RSS Difference	$RSS_{hysteresis}$	$RSS_{threshold}$
1	High	High	Lowest	Highest
2	High	Medium	Lower	Higher
3	High	Low	Low	High
4	Medium	High	Low	High
5	Medium	Medium	Medium	Medium
6	Medium	Low	High	Low
7	Low	High	Higher	Lower
8	Low	Medium	Higher	Lower
9	Low	Low	Highest	Lowest

Direction-biasing has been proposed in [60] to improve handoff performance. Austin [60] also shows that it is extremely difficult to estimate the direction of the MS with respect to the BSs at an intersection. Hence, direction-biasing can be utilized in an algorithm when there is sufficient confidence regarding direction estimates, and direction-biasing can be switched off when direction estimates are deemed unreliable. However, it is possible to exploit direction-biasing even when direction estimates are unreliable.

Figure 8.4 denotes the regions where direction estimates are unreliable by continuous lines and labels such regions as "No Direction Estimates" in the legend. Assume that the algorithm keeps track of reliable direction estimates and stores them as *previous direction estimates*. Consider the LOS handoff case when the MS travels from BS 0 to BS 2. The reliable direction estimates are available before the MS enters the intersection. According to these estimates, the MS is moving away from BS 0 and is approaching BSs 1, 2, and 3. Assume that the direction-biasing algorithm continues to use these estimates until new reliable direction estimates become available (after the MS traverses some distance beyond the intersection). After crossing the intersection, the MS is approaching BS 2, while moving away from BSs 0, 1, and 3. Thus, the direction estimates are wrong for BS 1 and BS 3 and correct for BSs 0 and 2. However, note that after the MS clears the intersection, the handoff candidates are LOS BSs, which are BS 0 and BS 2 in the case under consideration. Also note that the direction estimates being used by the algorithm are correct for both the good handoff candidate BSs (BS 0 and BS 2).

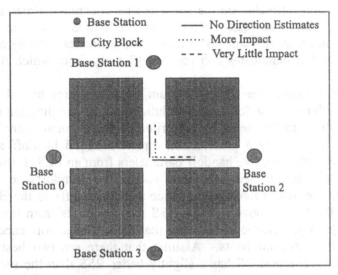

Figure 8.4: Direction-Biasing and Handoff Situations in a Microcellular System.

Thus, there is a very small region where the use of previous direction estimates can have an adverse impact on handoff performance. On the contrary, there is a relatively larger area where the use of previously reliable direction estimates may have no significant adverse impact on handoff performance. A similar situation exists for the NLOS handoff case. In the NLOS handoff case, the MS travels from BS 0 to BS 1. Again, the reliable direction estimates are available before the MS enters the intersection. According to these reliable estimates, the MS is receding from BS 0 and moving toward BSs 1, 2, and 3. Assume that the direction biasing algorithm continues to use these estimates until new reliable direction estimates become available (after the MS clears the intersection). After crossing the intersection, the MS is approaching BS 1, while moving away from BSs 0, 2, and 3. Thus, the direction estimates are wrong for BS 2 and BS 3 and correct for BS 0 and 1. However, note that after the MS clears the intersection, the handoff candidates are LOS BSs, which are BS 1 and BS 3. Also note that the direction estimates being used by the algorithm are correct for the BS being approached by the MS (BS 1). Moreover, the direction estimate for BS 0 (away from which the MS is moving) is also correct. There is a small region where the use of previously reliable direction estimates can adversely affect handoff performance, but there is a relatively large area where handoff performance does not suffer due to the use of previously reliable direction estimates. Thus, *it is more advantageous to use previous direction estimates than to switch off direction-biasing completely when the direction estimates are unreliable.* Note that the benefit of using previous direction estimates will be more pronounced for the LOS case than the NLOS case. In the LOS case, previous direction estimates are the same as true direction estimates for both good handoff candidates (LOS BSs). In the NLOS case, the previous direction estimates are the same as true direction estimates for only one of the good LOS handoff candidates and for the BS away from which the MS is moving.

Figure 8.5 shows the block diagram of a direction-biased adaptive handoff algorithm suitable for a microcellular system. The diagram is similar to the block diagram for the earlier algorithm, and the conventional handoff algorithm is replaced by a preselection direction-biased handoff algorithm [60] that accepts the adaptive handoff parameters from an FLS. Basically, the preselection direction-biased algorithm encourages handoffs to the BSs toward which the MS is moving and discourages handoffs to the BSs away from which the MS is moving. The RSS measurements from the BSs are biased through a preselection procedure that selects a handoff candidate by biasing the RSS measurements. Assume that there are two best handoff candidate BSs. The best BS has a slightly better RSS than the second best BS, and the MS is moving away from it. The MS is moving toward the second best BS. The preselection procedure ensures that the second best BS is preferred to the best BS since the second BS is more likely to be selected in the future than the currently best BS.

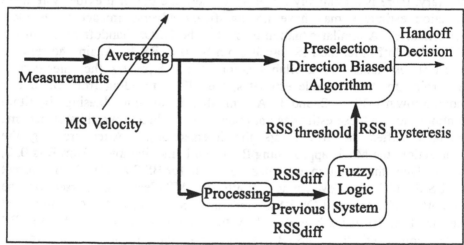

Figure 8.5: Block Diagram of a Direction-Biased Adaptive Microcellular Handoff Algorithm.

8.2.2 A Microcellular Algorithm with Interference, Traffic, and Mobility Adaptation

Figure 8.6 shows the block diagram of a handoff algorithm that uses a secondary FLS to provide interference, traffic, and mobility adaptation. This algorithm considers secondary handoff requirements whenever primary microcellular handoff objectives are not compromised. When an MS is near an intersection, there is a possibility of NLOS handoff, and any handoff parameter adaptation to obtain better performance in meeting secondary handoff objectives can adversely affect performance in meeting primary handoff objectives. Hence, the proposed algorithm switches on the secondary adaptation mechanism only under LOS handoff type situations (i.e., when an MS is relatively far from an intersection). The vicinity of an MS to an intersection can be predicted based on the RSS difference between best BSs or the reliability of direction estimates.

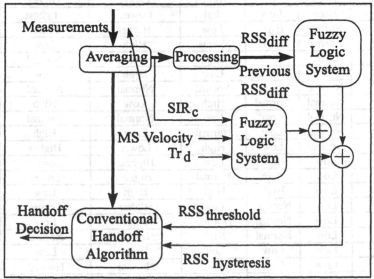

Figure 8.6: A Microcellular Handoff Algorithm With Traffic and Mobility Adaptation.

The primary FLS provides base values of the handoff parameters, while the secondary FLS provides incremental variations in $RSS_{threshold}$ and $RSS_{hysteresis}$ to reflect the dynamics of traffic, interference, and mobility (Table 8.2). The inputs to the secondary FLS are SIR_c, $Tr_d = Tr_c - Tr_n$, and *MS velocity*, and the outputs of the secondary FLS are incremental $RSS_{threshold}$ ($\Delta RSS_{threshold}$) and incremental $RSS_{hysteresis}$ ($\Delta RSS_{hysteresis}$). SIR_c is SIR of the current BS, and Tr_d is traffic difference (i.e., the difference in the number of calls in the current and the neighboring BS, $Tr_c - Tr_n$). *MS velocity* as an input to the FLS is the component of the *MS velocity* toward the serving BS. If the MS is moving toward the serving BS, the velocity is considered positive, and if the MS is moving away from the serving BS, the velocity is considered negative.

Table 8.2: Secondary Rule Base for a Microcellular Algorithm.

Rule Number	SIR_c	Tr_d	MS Velocity	$\Delta RSS_{threshold}$	$\Delta RSS_{hysteresis}$
1	High	High	Low	High	Low
2	High	High	Normal	Normal	Normal
3	High	High	High	Low	High
4	High	Normal	Low	Normal	Normal
5	High	Normal	Normal	Low	High
6	High	Normal	High	Lower	Higher
7	High	Low	Low	Low	High
8	High	Low	Normal	Lower	Higher
9	High	Low	High	Lowest	Highest
10	Normal	High	Low	Higher	Lower
11	Normal	High	Normal	High	Low
12	Normal	High	High	Normal	Normal
13	Normal	Normal	Low	High	Low
14	Normal	Normal	Normal	Normal	Normal
15	Normal	Normal	High	Low	High
16	Normal	Low	Low	Normal	Normal
17	Normal	Low	Normal	Low	High
18	Normal	Low	High	Lower	Higher
19	Low	High	Low	Highest	Lowest
20	Low	High	Normal	Higher	Lower
21	Low	High	High	High	Low
22	Low	Normal	Low	Higher	Lower
23	Low	Normal	Normal	High	Low
24	Low	Normal	High	Normal	Normal
25	Low	Low	Low	High	Low
26	Low	Low	Normal	Normal	Normal
27	Low	Low	High	Low	High

When all the inputs suggest a change in the handoff parameters in the same direction (i.e., either increase or decrease), the parameters are changed to the maximum extent. For example, consider Rule 9. "High" SIR_c indicates that the quality of the current link is very good. "Low" Tr_d indicates that there are very few users in the current cell. "High" *MS velocity* indicates that the MS is moving toward the current BS at a high speed. All these secondary FLS inputs suggest that handoff from the current BS be discouraged. Hence, the FLS makes $\Delta RSS_{threshold}$ "Lowest" (making overall $RSS_{threshold}$ smaller) and $\Delta RSS_{hysteresis}$ "Highest" (making overall $RSS_{hysteresis}$ large).

8.3 SIMULATION RESULTS

The simulation model used for the evaluation of microcellular algorithms is described in Chapter 3. The conventional algorithm has $RSS_{threshold}$, $RSS_{hysteresis}$, and $SIR_{threshold}$ as handoff parameters. $RSS_{threshold}$ is set as the RSS at the intersection from a BS. Setting $RSS_{hysteresis}$ as 7.5 dB gives a good compromise between the LOS and NLOS handoff performance with good cross-over distance for the NLOS case and a reasonable number of handoffs for the LOS case. $SIR_{threshold}$ is chosen to be 28 dB. For the non-direction-biased algorithm, the center of the input membership function for the set "Medium" is $RSS\_{diff(nom)}$ = 3.5 dB, and the centers of the input membership functions for the sets "Low" and "High" are located at the distance of ΔRSS_{diff} = 3.5 dB from $RSS_{diff(nom)}$. The center of the output membership function "Medium" for the fuzzy variable $RSS_{hysteresis}$ is located at $RSS_{hysteresis(nom)}$ = 7.5 dB, and the centers of the output membership functions for the extreme sets (e.g., "Highest" and "Lowest") are located at the distance of $\Delta RSS_{hysteresis}$ = 1.8 dB from $RSS_{hysteresis(nom)}$. The spreads of the membership functions are chosen in such a way that the membership value drops to zero at the center of the membership function of the nearest set. For the direction-biased algorithms, the hysteresis bias is $dir_{hysteresis}$ = 1 dB and the preselection direction-bias for RSS is hp_{rss} = 1.5 dB. A direction-biased algorithm that switches off direction biasing when the direction estimates are unreliable is referred to as a *restricted direction-biased algorithm*. A direction-biased algorithm that uses previous direction estimates when the direction estimates are unreliable is referred to as a *modified direction-biased algorithm*.

The performance of the algorithms is compared using the *operating point* as a performance metric as explained in Chapter 4. The cross-over distance should be as close as possible to the intersection (i.e., 255 m for the simulation model under consideration). It is assumed that the parameters of the conventional algorithm gives a good cross-over distance for the NLOS case, and, hence, the proposed algorithms should try to reduce the number of handoffs as much as possible while keeping the cross-over distance the same (or closer to the intersection).

8.3.1 LOS And NLOS Performance Evaluation of the Microcellular Algorithms

Table 8.3 shows the LOS operating points for the conventional and proposed non-direction-biased microcellular algorithms. As expected, the proposed algorithm reduces the number of handoffs since higher hysteresis values are used except near the intersection. The direction-biasing helps reduce the cross-over distance from 284.90 m to 279.40 m, an improvement of 5.50 m over the non-direction-biased basic algorithm. The number of handoffs is almost the same as for non-direction-biased and restricted direction-biased algorithms. The modified direction-biasing helps reduce the cross-over distance from 279.40 m to 268.22 m, an improvement of 11.18 m over the restricted direction-biased basic algorithm.

Table 8.3: LOS and NLOS Operating Points for Microcellular Algorithms.

Case	Conventional Algorithm	Adaptive Non-Direction-Biased Algorithm	Restricted Direction-Biased Algorithm	Modified Direction-Biased Algorithm
LOS	(268.2, 2.74)	(284.9, 1.66)	(279.4, 1.67)	(268.2, 1.72)
NLOS	(267.1, 3.45)	(269.3, 2.32)	(267.1, 2.35)	(267.1, 2.38)

The proposed non-direction-biased algorithm reduces the number of handoffs for the NLOS case. The tendency of the proposed algorithm to use relatively higher hysteresis values increases in the cross-over distance. However, direction-biasing will help reduce the cross-over distance, and, therefore, the non-direction-biased algorithm focuses on reducing the number of handoffs. The direction-biasing helps reduce the cross-over distance from 269.34 m to 267.11 m, an improvement of 2.23 m over the adaptive non-direction-biased algorithm for the NLOS case. The modified direction-biasing gives almost the same performance as the restricted direction-biasing.

8.3.2   Performance Evaluation of the Microcellular Algorithm with Interference, Traffic, and Mobility Adaptation

Table 8.4 summarizes the operating points for the LOS and the NLOS cases for the proposed modified direction-biased algorithm and the conventional algorithm.

Table 8.4: Operating Points for An Adaptive Microcellular Algorithm.

Case	Conventional Algorithm	Adaptive Modified Direction-Biased Algorithm
LOS	(268.2, 2.74)	(268.2, 1.72)
NLOS	(267.1, 3.45)	(267.1, 2.38)

The number of handoffs is reduced by 1.02 (or 37%) for the LOS case and 1.07 (or 31%) for the NLOS case, while the cross-over distances are preserved for both LOS and NLOS cases. Note that if the same number of handoffs were to be obtained for the conventional algorithm for LOS and NLOS cases, it would have required larger hysteresis, leading to higher cross-over distance, increasing the interference in the LOS case and increasing the call drop probability (due to insufficient RSS) for the NLOS case. Thus, the proposed algorithm is a good generic algorithm suitable for a microcellular environment. This section uses the modified direction-biased algorithm as a basic algorithm and uses a secondary mechanism (an FLS) to obtain adaptation to interference, traffic, and mobility. Adaptation to interference and traffic can be judged based on the distribution of SIR and the number of active users in a cell, respectively. The mobility adaptation can be analyzed using cross-over distance as a performance metric. This section shows the advantages of incorporating interference, traffic, and mobility adaptation into the modified direction-biased algorithm.

Figure 8.7 shows the RSS distribution for the conventional and proposed adaptive algorithms for a LOS handoff scenario. As expected, there is a degradation in the CDF of the RSS (maximum of 0.4 dB) for the proposed algorithm since the direction-biasing tends to use the BSs that have the potential of being selected in the future and not the strongest RSS BSs at the present time.

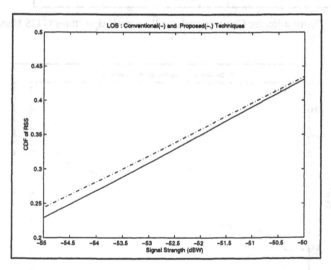

Figure 8.7: RSS Distribution for Conventional and Proposed Algorithms (LOS Handoff).

Figure 8.8 shows the SIR distribution for the conventional and proposed adaptive algorithms for a LOS handoff scenario. There is an improvement of 0.5 dB in the SIR distribution for the proposed algorithm. This indicates that the proposed algorithm is adaptive to interference, and BSs with potentially better quality (quantified by higher SIR) are preferred. Figure 8.9 shows the traffic distribution for the conventional and proposed adaptive algorithms for a LOS handoff scenario. The traffic adaptation improves the traffic distribution by 0.25 calls compared to the conventional algorithm. The proposed algorithm allows relatively more users to be served by the system due to this traffic balancing.

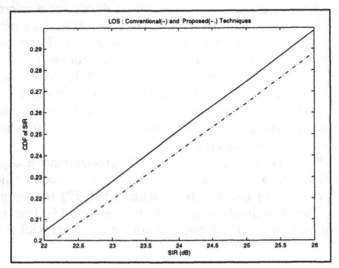

Figure 8.8: SIR Distribution for Conventional and Proposed Algorithms (LOS Handoff).

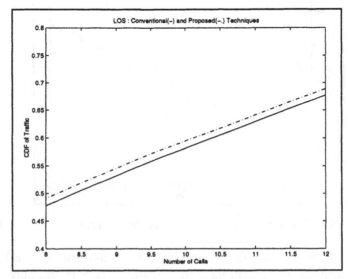

Figure 8.9: Traffic Distribution for Conventional and Proposed Algorithms (LOS Handoff).

Figure 8.10 shows the RSS distribution for the conventional and proposed adaptive algorithms for a NLOS handoff scenario. There is a maximum degradation of about 0.3 dB in the RSS distribution for the proposed algorithm due to direction-biasing. Figure 8.11 shows the SIR distribution for the conventional and proposed adaptive algorithms for a NLOS handoff scenario. There is an improvement of 0.5 dB in the SIR performance of the proposed algorithm.

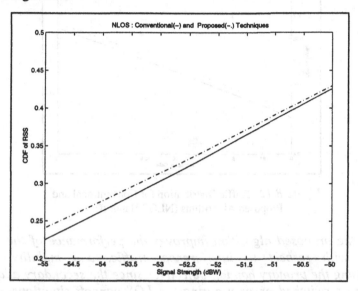

Figure 8.10: RSS Distribution for Conventional and Proposed Algorithms (NLOS Handoff).

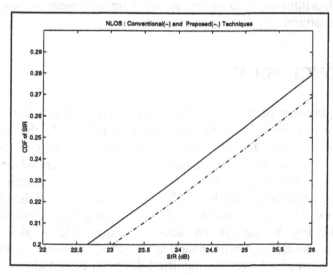

Figure 8.11: SIR Distribution for Conventional and Proposed Algorithms (NLOS Handoff).

Figure 8.12 shows the traffic distribution for the conventional and proposed adaptive algorithms for a NLOS handoff scenario. The traffic adaptation improves the traffic distribution by 0.25 calls compared to the conventional algorithm.

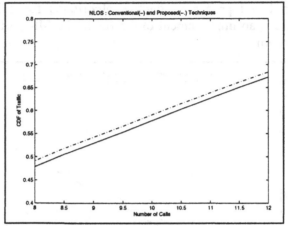

Figure 8.12: Traffic Distribution for Conventional and
Proposed Algorithms (NLOS Handoff).

Note that the proposed algorithm improves the performance of the handoff algorithm with respect to interference, traffic, and mobility without compromising the primary handoff objectives since the secondary adaptation mechanism is switched on only during the LOS handoff situations. Hence, there is not a significant improvement due to adaptation. Nevertheless, the proposed algorithm tends to obtain as much improvement as possible under the given constraints.

8.4   CONCLUSION

The deployment of microcells increases system capacity but complicates resource management due to the characteristics of the propagation environment. This chapter proposes handoff algorithms that perform uniformly well in different handoff scenarios. Knowledge of the microcellular environment is utilized to design fuzzy logic systems that render the algorithms adaptation capability. The proposed modified direction biased algorithm exploits characteristics of the propagation environment and direction biasing to design an adaptive microcellular algorithm. A microcellular handoff algorithm adaptive to traffic, interference, and mobility has been superimposed over the modified direction biased algorithm to meet both the primary and secondary handoff objectives.

Chapter 9

OVERLAY HANDOFF ALGORITHMS

An overlay system is a hierarchical architecture that uses large macrocells to overlay clusters of small microcells. The overlay system attempts to balance maximizing system capacity and minimizing cost and when done correctly provides much flexibility to handle changing demands for a region or distinctly different traffic patterns in a region. For example, fast moving traffic would require many handoffs if handled by the microcells. Furthermore, additional channels for a region are needed during certain times (e.g., rush hour), while at another time of the day, these channels are needed for a different region. This chapter proposes an adaptive overlay handoff algorithm that improves overall system performance.

9.1 INTRODUCTION TO HANDOFFS IN OVERLAYS

An overlay system consists of large macrocells and small microcells. The macrocells are also referred to as *overlay cells* and overlay a cluster of microcells. Figure 9.1 shows a macrocell-microcell overlay system and four generic handoff scenarios. Cell 1 and Cell 2 are macrocells that overlay clusters of microcells. A cluster of microcells consists of cells A, B, C, and D. Four generic types of handoffs are *macrocell to macrocell*, *macrocell to microcell*, *microcell to microcell*, and *microcell to macrocell*. When an MS travels from one macrocell to another (Case 1), a macrocell to macrocell handoff occurs. This type of handoff typically occurs near the macrocell borders. When an MS enters a microcell from a macrocell (Case 2), a macrocell to microcell handoff occurs. Even though the signal strength from the macrocell is usually greater than the signal strength from the microcell (due to relatively higher macrocell BS transmit power), this type of handoff is made to utilize the microcell connection that is economical, power efficient, and spectrally efficient in that it generates less interference than a macrocell. When an MS leaves a microcell (Case 3) and enters a macrocell, a microcell to macrocell handoff is made to save the call since the microcell can no longer provide a good quality communication link to the MS. When an MS travels from one microcell to another (Case 4), a microcell to microcell handoff is made to reduce power requirements and get a better quality signal. An overlay system achieves a balance between maximizing the system capacity and minimizing the cost. Microcells cover areas with high traffic intensities while macrocells provide wide area coverage. Small cells can provide very high capacity but lead to an expensive system due to infrastructure costs. Important considerations for designing efficient handoff algorithms for overlay systems are outlined below.

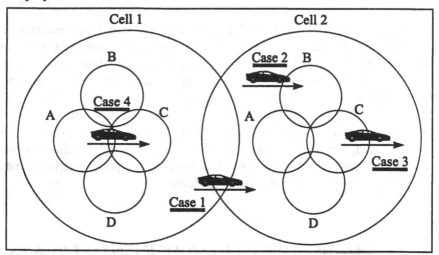

Figure 9.1: Generic Handoff Scenarios in a Macrocell-Microcell Overlay System.

- **Service**. An attempt should be made to maximize the microcell usage since the microcell connection has a low cost due to the better frequency reuse factor and low transmit power requirements. However, far regions should be served by macrocells for a better quality communication link. Microcell overflow traffic should be handled by macrocells.

- **Mobility**. High-speed vehicles should be connected to macrocells to reduce the handoff rate and the associated network load. This will also enable a handoff algorithm to perform uniformly well for LOS and NLOS handoffs in microcells. The handoff parameters can now be optimized for LOS handoff situations since the requirement of a very fast handoff for a typical NLOS situation can be easily avoided by connecting high-speed users to macrocells.

- **Propagation Environment**. In an overlay system, a user experiences both macrocell and microcell environments as the user travels across macrocells and microcells. Different fading intensities (e.g., low and high) exist in macrocells and microcells.

- **Resource Management**. Resource management in an overlay system is a difficult task. One of the crucial issues is an optimum distribution of channels between macrocells and microcells.

- **Specific Handoff Requirements**. A handoff algorithm should perform uniformly well in the four generic handoff situations described earlier. The algorithm should attempt to achieve the goal of an overlay system (i.e., the balance between the microcell usage and network load). The algorithm should balance traffic in the cells.

An urban cellular system that has a single macrocell overlaying four clusters of microcell with each cluster having four microcells is considered in [73]. A tradeoff between the network capacity and the probability of handoff failure has been examined. The roles of queuing of handoff requests at the microcell level and channel reservation at the macrocell level are investigated. Lin [98] proposes an analytical model to study the performance of a PCS overlay system. An iterative algorithm computes the overflow traffic from a microcell to macrocell and uses this traffic measure to compute the call completion probability. The study shows that the variance of the microcell residual time distribution and the number of microcells covered by a macrocell significantly affect the call completion probability. Lagrange [100] proposes call admission and handoff strategies for an overlay system. Macrocells accept handoff requests that cannot be entertained by microcells. A handoff criterion that is sensitive to the mobile's speed is proposed in [32] to differentiate between slow and fast users in an overlay system. Handoffs from macrocells to microcells are avoided for fast moving vehicles while slow users are dropped from macrocells and connected to microcells. The paper shows that handoff rate can be reduced significantly using the speed sensitive handoff criterion. Furukawa [30] exploits a self-organized dynamic channel assignment and automatic transmit power control to obviate the need for their redesigning. The available channels are reused between macrocells and microcells. A slight increase in transmit power for the microcells compensates for the interference from macrocell to microcell.

A new adaptive handoff algorithm suitable for overlay systems is proposed and is described in Section 9.2. Section 9.3 analyzes the performance of the algorithms from various aspects. Finally, Section 9.4 summarizes the chapter.

9.2 CONVENTIONAL AND ADAPTIVE OVERLAY HANDOFF ALGORITHMS

This section describes a conventional overlay algorithm and a proposed adaptive algorithm. The objective of this adaptive algorithm is to meet the handoff objectives of an overlay system by adapting handoff parameters. Figure 9.2 shows the block diagram of one possible conventional algorithm for an overlay system. Handoff criteria are averaged according to a velocity adaptive averaging mechanism. The *Handoff Initiation Mechanism* compares the RSS of the current BS (RSS_c) with a fixed threshold (RSS_{th}). If the current BS cannot provide sufficient RSS, the handoff process is initiated. The *Cell Selection Mechanism* determines the best macrocell "x" and the best microcell "y" as potential handoff candidates. The *Handoff Decision Mechanism* is explained next.

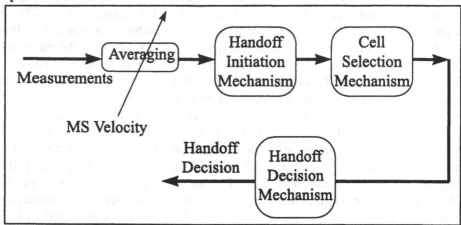

Figure 9.2: Block Diagram of a Conventional Overlay Handoff Algorithm.

If the currently serving cell is a macrocell, the sequence of Figure 9.3 is followed, and if the currently serving cell is a microcell, the sequence of Figure 9.4 is followed.

- If the current BS is a macrocell, the speed of the MS may be high or low. Hence, based on the speed, there is a different sequence for a handoff decision. If the MS speed is greater than the velocity threshold V_{th}, RSS_c is compared with RSS_x and a handoff is made to the macrocell "x" if RSS_x exceeds RSS_c by an amount RSS_{hyst} (hysteresis value). If the MS speed is less than V_{th}, the first attempt is made to connect the user to the microcell "y." If RSS_y exceeds the absolute threshold RSS_{th} by an amount RSS_{hyst}, a handoff is made to the microcell "y." However, if the microcell BS cannot satisfy this condition, an attempt is made for a potentially better macrocell connection. If RSS_x exceeds RSS_c by an amount RSS_{hyst}, a handoff is made to the macrocell "x."

- If the current BS is a microcell, the MS speed is low and the speed does not dictate a specific sequence of steps. If RSS_y exceeds the threshold RSS_{th} by an amount RSS_{hyst}, a handoff is made to the microcell "y." However, if the microcell BS cannot meet this requirement, an attempt is made for the macrocell BS. If RSS_x exceeds RSS_{th} by an amount RSS_{hyst}, handoff is made to the macrocell "x."

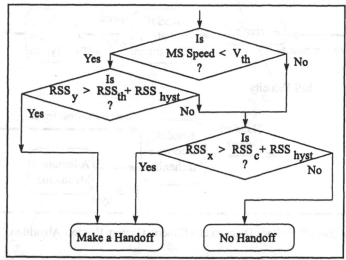

Figure 9.3: The Sequence of Steps for a Current Macrocell Connection.

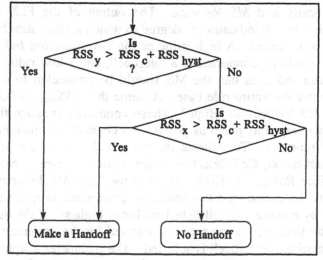

Figure 9.4: The Sequence of Steps for a Current Microcell Connection.

Figure 9.5 shows the block diagram of the proposed generic algorithm for an overlay system. Handoff criteria are averaged according to a velocity adaptive averaging mechanism. The *Handoff Initiation Mechanism* compares the RSS of the current BS (RSS_c) to a fixed threshold (RSS_{th}). If the current BS cannot provide sufficient RSS, the handoff process is initiated. An FLS serves as a *Cell Selection Mechanism* to determine the best macrocell "x" and the best microcell "y" as potential handoff candidates.

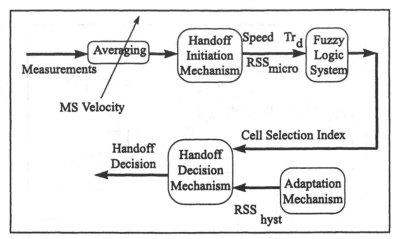

Figure 9.5: Block Diagram of a Generic Overlay Handoff Algorithm.

The inputs of the FLS are RSS_{micro} (RSS from the microcell BS), Tr_d (traffic difference or the difference in the number of calls in the microcell and in the macrocell), and *MS Velocity*. The output of the FLS is the *Cell Selection Index*, which indicates the degree to which a given user belongs to a microcell or a macrocell. A high value of the *Cell Selection Index* indicates that the MS should be connected to a microcell, and a low value of the *Cell Selection Index* indicates that the MS should be connected to a macrocell. Table 9.1 shows the entire rule base. Assume that RSS_{micro} is "Low," Tr_d is "High," and *MS Velocity* is "High." These conditions indicate that the call should be encouraged as much as possible to connect to a macrocell; this is rule number nineteen. To indicate the highest degree of confidence for a macrocell connection, Cell Selection Index is made lowest. Now consider rule nine. Since RSS_{micro} is "High," Tr_d is "Low," and *MS Velocity* is "Low," the call should be encouraged to connect to a microcell as much as possible. This is done by making the Cell Selection Index highest. If the output of the FLS is greater than zero, a microcell is selected for the communication with the MS; otherwise, a macrocell is selected. The parameter adaptation for the proposed generic overlay algorithm is explained next.

- If the currently serving cell is a macrocell and the candidate cell is also a macrocell, an incremental hysteresis $\Delta hyst_{macro}$ is found as shown in Figure 9.6. If the MS is moving toward both the current and the candidate BSs or moving away from the BSs, a fixed hysteresis h_{macro} is used. If the MS is moving toward the current BS and moving away from the candidate BS, a handoff is discouraged by increasing the hysteresis value by an amount Δh_{macro}. However, if the MS is moving toward the candidate BS and away from the current BS, handoff is encouraged by decreasing the hysteresis value by an amount Δh_{macro}. The overall adaptive hysteresis is $RSS_{effective} = RSS_{hyst} + \Delta hyst_{macro}$ where RSS_{hyst} is the nominal value of the RSS hysteresis. This adaptation of hysteresis is based on direction biasing and helps reduce the ping-pong effect between two macrocells. A handoff is made to the macrocell "x" if RSS_x exceeds RSS_c by an amount $RSS_{effective}$.

Table 9.1: Rule Base for Cell Selection.

Rule Number	RSS_{miscro}	Tr_d	MS Velocity	Cell Selection Index
1	High	High	High	Low
2	High	High	Normal	Normal
3	High	High	Low	High
4	High	Normal	High	Normal
5	High	Normal	Normal	High
6	High	Normal	Low	Higher
7	High	Low	High	High
8	High	Low	Normal	Higher
9	High	Low	Low	Highest
10	Normal	High	High	Lower
11	Normal	High	Normal	Low
12	Normal	High	Low	Normal
13	Normal	Normal	High	Low
14	Normal	Normal	Normal	Normal
15	Normal	Normal	Low	High
16	Normal	Low	High	Normal
17	Normal	Low	Normal	High
18	Normal	Low	Low	Higher
19	Low	High	High	Lowest
20	Low	High	Normal	Lower
21	Low	High	Low	Low
22	Low	Normal	High	Lower
23	Low	Normal	Normal	Low
24	Low	Normal	Low	Normal
25	Low	Low	High	Low
26	Low	Low	Normal	Normal
27	Low	Low	Low	High

- If the currently serving cell is a macrocell and the candidate cell is a microcell, an incremental hysteresis is taken as Δh_{micro}. The overall adaptive hysteresis is $RSS_{effective} = RSS_{hyst} - \Delta h_{micro}$. This adaptation of hysteresis is based on the idea of encouraging handoffs to the microcells to increase microcell usage. A handoff is made to the microcell "y" if RSS_y exceeds RSS_{th} by an amount $RSS_{effective}$.

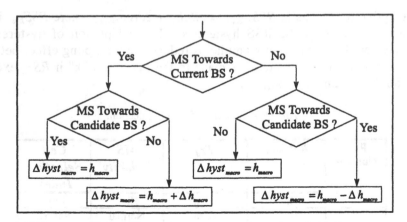

Figure 9.6: Adaptive Handoff Parameters for a Current Macrocell Connection.

- If the currently serving cell is a microcell and the candidate cell is also a microcell, an incremental hysteresis $\Delta hyst_{micro}$ is found as shown in Figure 9.7. If the MS is moving toward the current BS and away from the candidate BS, a handoff is discouraged by increasing the hysteresis value by an amount Δh_{micro}. However, if the MS is moving toward the candidate BS and away from the current BS, a handoff is encouraged by decreasing the hysteresis value by an amount Δh_{micro}. The overall adaptive hysteresis is $RSS_{effective} = RSS_{hyst} + \Delta hyst_{micro}$. This adaptation of hysteresis is based on direction biasing and helps reduce the ping-pong effect between two microcells. A handoff is made to the microcell "y" if RSS_y exceeds RSS_c by an amount $RSS_{effective}$.

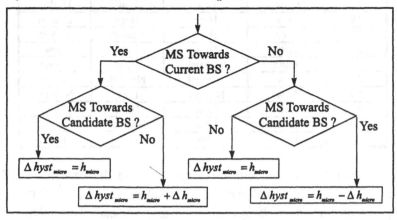

Figure 9.7: Adaptive Handoff Parameters for a Current Microcell Connection.

- If the currently serving cell is a microcell and the candidate cell is a macrocell, an incremental hysteresis is taken as Δh_{macro}. The overall hysteresis is $RSS_{effective} = RSS_{hyst} - \Delta h_{macro}$. This adaptation of hysteresis is based on the idea of encouraging handoffs to the macrocells from microcells to save the call since the microcell coverage area is limited and since the call may be dropped if handoff is not made early enough. A handoff is made to the macrocell "x" if RSS_x exceeds RSS_{th} by an amount $RSS_{effective}$.

9.3 SIMULATION RESULTS FOR OVERLAY ALGORITHMS

The simulation model used to evaluate overlay handoff algorithms is described in Chapter 3. Table 9.2 lists the parameters of the algorithms and simulation parameters.

Table 9.2: Simulation and Algorithm Parameters.

Parameter	Value
Macrocell Radius	2.5 km
Microcell Radius	700 m
Macrocell BS Transmit Power	290 W
Microcell BS Transmit Power	0.1 mW
No. of Channels/BS	16
Mean Call Duration	120
Normalized Traffic Load	0.7
h_{macro}	4 dB
Δh_{macro}	4 dB
h_{micro}	2 dB
Δh_{micro}	1 dB
Velocity Threshold (V_{th})	45 mph

The center of the input membership function for the set "Normal" of the fuzzy variable RSS_{micro} is $RSS_{nom} = RSS_{mid} - 7\text{dB}$ (RSS_{mid} is the power received at the boundary of a microcell in the absence of fading), and the centers of the input membership functions for the sets "Low" and "High" are located at the distance of $\Delta RSS = 7$ dB from RSS_{nom}. The center of the input membership function for the set "Normal" of the fuzzy variable Tr_d is $Tr_{d(nom)} = 0$, and the centers of the input membership functions for the sets "Low" and "High" are located at the distance of two from $Tr_{d(nom)}$. The center of the input membership function for the set "Normal" of the fuzzy variable *MS Velocity* is

$V_{nom} = 45$ mph, and the centers of the input membership functions for the sets "Low" and "High" are located at the distance of $\Delta V = 25$ from V_{nom}. The center of the output membership function "Normal" for the fuzzy variable Cell Selection Index is located at $Index_{nom} = 0$, and the centers of the output membership functions for the extreme sets (e.g., "Highest" and "Lowest") are located at the distance of $\Delta Index = 3$ from $Index_{nom}$. The spreads of the membership functions are chosen in such a way that the membership value drops to zero at the center of the membership function of the nearest set.

Figure 9.8 shows the CDF of RSS for the conventional and proposed algorithm. The proposed algorithm improves the RSS distribution that can be attributed to the better Cell Selection Mechanism. Figure 9.9 shows the CDF of SIR for the conventional and proposed algorithm. The proposed algorithm improves the SIR distribution, leading to better quality communication links.

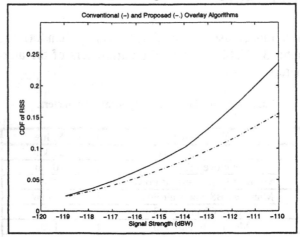

Figure 9.8: Distribution of RSS for the Conventional and Proposed Algorithms.

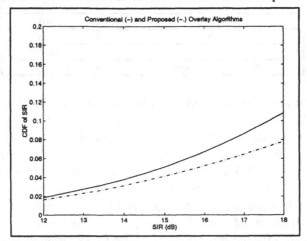

Figure 9.9: Distribution of SIR for the Conventional and Proposed Algorithms.

Figure 9.10 shows the traffic distribution for the conventional and proposed adaptive algorithms. The proposed algorithm improves the traffic distribution, balancing the traffic in the adjacent cells.

Figure 9.11 shows the microcell usage for the conventional and proposed adaptive algorithms as a function of time. A snapshot of the simulation for a time window from 300 sec to 500 sec is shown. The microcell usage factor is defined as the fraction of the total number of calls connected to microcells at a given instant. For example, if all the users are connected to microcells at a given instant, the microcell usage factor is one. If all the users are connected to macrocells at a given instant, the microcell usage factor is zero.

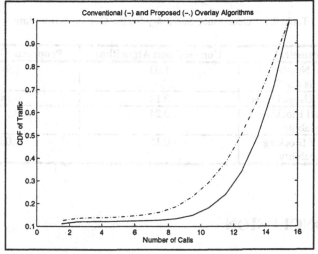

Figure 9.10: Traffic Distribution for the Conventional and Proposed Algorithms.

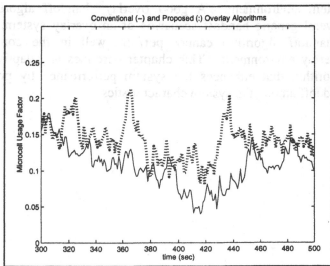

Figure 9.11: Microcell Usage for the Conventional and Proposed Algorithms.

Table 9.3 summarizes performance of the algorithms. The proposed algorithm tends to increase microcell usage as desired. This shows the importance of adaptive parameters since the fixed parameter algorithm restricts the microcell usage by the limitation of fixed parameters. The number of handoffs per call increases due to increased microcell usage. Note that the RSS performance is not compromised by the microcell connections (evident from the improvement in RSS performance). Traffic balancing reduces the call blocking probability by a factor of 1.8 and handoff blocking probability by a factor of 3.

Table 9.3: Performance Summary for the Overlay Algorithms.

Parameter	Conventional Algorithm	Proposed Algorithm
Average Number of Handoffs per Call	1.03	1.94
Microcell Usage Factor	0.12	0.14
New Call Blocking Probability	0.24	0.13
Handoff Blocking Probability	0.15	0.05

9.4   CONCLUSION

Resource management tasks in the overlay system are complicated by the overlay system environment. A good overlay handoff algorithm must consider several generic handoff scenarios of an overlay system. A fixed parameter handoff algorithm cannot perform well in the complex and dynamic overlay environment. This chapter describes an adaptive overlay handoff algorithm that enhances the system performance by providing a balanced tradeoff among the system characteristics.

Chapter 10

SOFT HANDOFF ALGORITHMS

Soft handoff is an approach in which the next connection is made before the current connection is broken. Soft handoff exploits spatial diversity attributed to signals emanating from multiple base stations to increase received signal energy for more reliable handoff performance. A good soft handoff algorithm achieves a balance between the quality of the received signal and the associated cost in system resources. This chapter highlights important considerations for soft handoff and develops adaptation mechanisms for new soft handoff algorithms. Specifically, two new soft handoff algorithms that provide high performance by adapting to the dynamic cellular environment are proposed and evaluated.

10.1 INTRODUCTION TO SOFT HANDOFF

An MS in soft handoff communicates simultaneously with more than one BS. Soft handoff exploits spatial diversity to increase overall signal energy for better performance. The CDMA technology implements soft handoff. Figure 10.1 shows generic soft handoff scenarios in a cellular system. When an MS is close to a BS (e.g., near BS C), it communicates with only that BS (BS C). However, near cell borders, it is relatively far from all adjacent BSs, and the RSS from a single BS may not be sufficient to provide a good quality communication link. In such a case, the MS combines signals from multiple BSs to obtain a good quality signal using one of the diversity combining techniques such as equal gain combining, selection combining, and maximal ratio combining. For example, in the overlap region between Cell A and Cell D, the MS is connected to both BS A and BS D, leading to a *two-way soft handoff* scenario. There may also be a *three-way soft handoff* in which the MS communicates with three BSs (e.g., BSs A, B, and C in Figure 10.1). When sectorization is employed, the MS communicates with one or more sectors from multiple BSs.

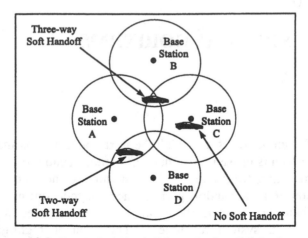

Figure 10.1: Generic Soft Handoff Scenarios in a Cellular System.

A glimpse of research work on soft handoff is provided next. Salmasi [21] discusses system design aspects of CDMA including signal and waveform design, power control, soft handoff, and variable data rates. Simmonds [85] explains the concepts of macroscopic diversity and soft handoff. Results are reported for propagation studies that evaluate soft handoff performance using a wideband correlation type channel sounding system. The advantages of soft handoff are also highlighted. Wang [87] presents simulation results on the effects of soft handoff, frequency reuse, and non-ideal antenna sectorization on CDMA system capacity. The simulation results provide statistics of soft and softer handoffs for different values of handoff parameters. The simulation model consists of a cell layout of nineteen hexagonal cells with either omni-directional or three sectored antennas. The mobiles are uniformly distributed in the service region. The propagation model consists of log-linear path loss with uncorrelated log-normal shadowing. The pilot powers received at the MS from the BSs are ranked according to the signal strengths and compared with a handoff threshold. The number of pilots, K, with power greater than a handoff threshold is counted, and the distribution of K is found as a function of handoff threshold. It is recommended that the MS have at least three demodulators to efficiently combine the signals from the BSs. Zhang [107] analyzes the tradeoff between diversity exploitation and effective resource utilization in soft handoff. The analysis quantifies the handoff performance by the number of active set updates, the number of BSs involved in soft handoff, and the outage probability of the received signal. The analysis results are validated with a simulation model that consists of a cell layout with two BSs and the MS traveling from one BS to another at a constant velocity.

The propagation environment is characterized by log-linear path loss with correlated shadow fading.

Chheda [119] analyzes performance of the IS-95 soft handoff algorithm. A *simplified* soft handoff mechanism of IS-95 adapted from [119] is briefly summarized here. The IS-95 system implements MAHO (see Chapter 1). The MS measures pilot E_c/I_o of surrounding cells (or sectors). Each sector transmits a pilot signal with the same power. Hence, the ratio of the received pilot power and the out-of-cell interference plus thermal noise for a given sector is a good indicator of the signal quality from that sector. Note that this ratio is equivalent to the ratio of pilot chip energy to interference energy E_c/I_o. In other words, a high pilot E_c/I_o from a sector indicates that good signal quality signals will be received from that sector. The sectors with a sufficiently large pilot E_c/I_o (measured at the MS) form an Active Set (AS). The MS communicates with members of the AS on both forward and reverse links. In other words, members of the AS transmit the same information signal to the MS and receive the same information signal from the MS. On the forward link, the RAKE fingers of the MS demodulate signals from a maximum of three AS members. On the reverse link, the signals received at the AS sectors of the same cell are combined using maximal ratio combining, while the best signal is selected from multiple BSs that have their sectors as members of the AS (an example of the implementation of selection diversity). Each sector has two receive antennas at the cell site to exploit diversity. If the MS measures a pilot that has E_c/I_o greater than a threshold, T_ADD, it is made a member of the AS. If the pilot strength of an AS member remains below a threshold, T_DROP, for T_TDROP seconds, it is removed from the AS. Thus, sectors with pilots that are received with strong E_c/I_o at the MS remain in the AS. Such SHO management ensures that reasonably strong signals that can provide diversity gain are utilized.

Several factors impact the performance of soft handoff algorithms. A good soft handoff algorithm attempts to achieve a balance between the quality of the signal and the associated cost. In general, the greater the number of BSs involved in soft handoff, the better the quality of the signal due to increased diversity gain and the higher the degree to which the network resources are consumed. Important considerations for designing soft handoff algorithms are as follows.

- **Cellular System Layouts**. Soft handoff may be implemented in a macrocellular, microcellular, or overlay system. Traffic, mobility, and propagation environment in these distinct system deployment scenarios should be considered.
- **Primary Soft Handoff Requirements**. The handoff algorithm should try to maximize signal quality and minimize the number of BSs involved in soft handoff. The number of Active Set (i.e., the set that contains the list

of BSs in soft handoff) updates should be minimized to reduce the network load.

- **Secondary Soft Handoff Requirements**. The algorithm should be adaptive to correspond to vehicle speed. The algorithm should attempt to balance traffic among base stations.

This chapter proposes two adaptive soft handoff algorithms that consider primary and secondary soft handoff requirements (or objectives).

10.2 ADAPTIVE SOFT HANDOFF ALGORITHMS

The first of the two new algorithms attempts to satisfy primary handoff requirements. The second uses the first algorithm to meet primary handoff objectives and implements an adaptation mechanism to achieve the secondary handoff goals.

10.2.1 A Generic Soft Handoff Algorithm

Figure 10.2 shows the block diagram of a generic adaptive soft handoff algorithm. A velocity adaptive averaging mechanism is used to average handoff criteria (see Chapter 4). One possible conventional soft handoff algorithm is illustrated in Figure 10.3.

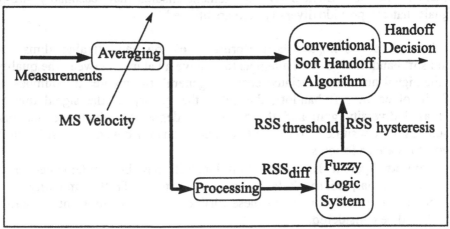

Figure 10.2: Block Diagram of an Adaptive Soft Handoff Algorithm.

The conventional algorithm has threshold ($RSS_{threshold}$) and hysteresis ($RSS_{hysteresis}$) as handoff parameters. To qualify as a member of the Active Set, the BS must pass two tests. Test 1 evaluates an absolute strength of the BS, and Test 2 compares the relative strength of the BS with respect to the best BS in the Active Set. Test 1 ensures that the BS, if admitted to the Active Set, can contribute significantly to the improvement in overall signal quality. Test 2 keeps only the best available BSs in the Active Set and tries to minimize resource utilization.

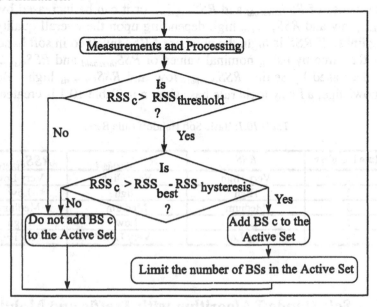

Figure 10.3: A Conventional Soft Handoff Algorithm.

According to Test 1, if RSS of a BS c (RSS_c) exceeds $RSS_{threshold}$, BS c is considered one of the potential soft handoff participants. According to Test 2, BS c actually becomes a member of the active set if RSS_c is greater than the RSS of the best BS in the active set (RSS_{best}) minus $RSS_{hysteresis}$. The RSS of each BS is evaluated using this comparison procedure, and a set of the BSs that pass these tests is determined. The number of BSs in the Active Set is limited to the number of demodulators, N, available at the MS. The Active Set consists of best N BSs if all of these N BSs have passed the tests explained earlier. The handoff parameters are adapted using an FLS. The input to the FLS is the RSS at the MS at the previous sample time. The outputs of the FLS are adaptive handoff parameters, $RSS_{threshold}$ and $RSS_{hysteresis}$. When an MS is relatively close to a BS, the RSS at the MS is very high, and there is no need to initiate soft handoff since the current communication link already has sufficiently high RSS. To discourage any BS from becoming a member of the current Active Set (that consists of only the current BS), $RSS_{threshold}$ is set very high and $RSS_{hysteresis}$ is set very low. When the MS is far from the neighboring BSs and the MS is connected to only one BS, RSS is very low, and, hence, soft handoff should be initiated to increase overall signal strength. Soft handoff can now be encouraged by setting $RSS_{threshold}$ very low and $RSS_{hysteresis}$ very high. When the MS is far from the neighboring BSs but the MS is connected to more than one BS, RSS may be high or low depending upon the number of BSs and the quality of existing MS-BS connections. If RSS is low, the number of BSs involved in soft handoff can be kept the same by using nominal values of $RSS_{threshold}$ and $RSS_{hysteresis}$, or it can be increased by setting $RSS_{threshold}$ low and $RSS_{hysteresis}$ high depending upon the overall quality of the existing links. If RSS is high, the number of BSs involved in soft handoff can be kept the same by using nominal values of $RSS_{threshold}$ and $RSS_{hysteresis}$ or, it can be decreased by setting $RSS_{threshold}$ low and $RSS_{hysteresis}$ high. Based on such knowledge, a fuzzy logic rule base shown in Table 10.1 is created.

Table 10.1: Basic Soft Handoff Rule Base.

Rule Number	RSS	$RSS_{threshold}$	$RSS_{hysteresis}$
1	Very High	Very High	Very Low
2	High	High	Low
3	Medium	Medium	Medium
4	Low	Low	High
5	Very Low	Very Low	Very High

10.2.2 A Soft Handoff Algorithm with Traffic and Mobility Adaptation

Figure 10.4 shows the block diagram of a handoff algorithm with traffic and mobility adaptation. This algorithm uses the primary FLS to provide adaptive handoff parameters to the conventional soft handoff algorithm and a secondary FLS to meet the secondary handoff goals. As discussed earlier, the input to the primary FLS is the RSS at the MS at the previous sample time. The outputs of the primary FLS are adaptive handoff parameters, $RSS_{threshold}$ and $RSS_{hysteresis}$. Inputs to the secondary FLS are Tr (traffic or number of users in a cell) and *MS velocity* (component of the velocity toward the BS). If the MS is moving toward a BS, the *MS velocity* is considered positive, and if the MS is moving away from the BS, the *MS velocity* is considered negative. The output of the secondary FLS is a preselection index that indicates the degree to which the BS is a good candidate for the Active Set if traffic and mobility were the only considerations.

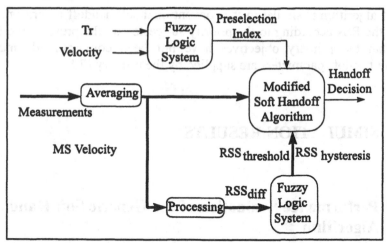

Figure 10.4: A Soft Handoff Algorithm with Traffic and Mobility Adaptation.

A complete fuzzy logic rule base is shown in Table 10.2. Consider Rule 7. The traffic in the cell is "Low" (meaning that there are very few users in the cell), and the velocity is "High" (meaning that the MS is moving at a high velocity toward the BS). This situation suggests that the BS under consideration should be encouraged to the maximum extent to become a member of the Active Set. This would help achieve traffic balancing and reduce the number of Active Set updates.

Table 10.2: Rule Base for Traffic and Mobility Adaptation.

Rule Number	Traffic (*Tr*)	*MS Velocity*	Preselection Index
1	High	High	Medium
2	High	Medium	Low
3	High	Low	Very Low
4	Medium	High	High
5	Medium	Medium	Medium
6	Medium	Low	Low
7	Low	High	Very High
8	Low	Medium	High
9	Low	Low	Medium

Now consider Rule 3. The traffic in the cell is "High" (meaning that there are many users in the cell), and the velocity is "Low" (meaning that the MS is moving at a high velocity away from the BS). This situation suggests that the BS under consideration should be discouraged to the maximum extent from becoming a member of the Active Set. This would again help achieve traffic balancing and reduce the number of Active Set updates. The modified soft handoff algorithm is similar to the conventional soft handoff algorithm, but it selects the BSs according to the priority suggested by the preselection index. Note that the primary objectives have not been compromised since the adaptive handoff parameters are supplied by the primary FLS.

10.3   SIMULATION RESULTS

10.3.1 Performance Evaluation of the Generic Soft Handoff Algorithm

The simulation model used to evaluate soft handoff algorithms is described in Chapter 3. The conventional algorithm has $RSS_{threshold}$ and $RSS_{hysteresis}$ as handoff parameters. $RSS_{threshold}$ is the RSS at the distance of cell radius (excluding the fading variations), and $RSS_{hysteresis} = 8$ dB. For the primary FLS, the center of the input membership function for the set "Medium" is RSS_{nom}, the same as $RSS_{threshold}$ for the conventional algorithm. The centers of the input membership functions for the sets "Low" and "High" are located at the distance of ΔRSS_{nom} dB from RSS_{nom} where ΔRSS_{nom} is the difference in the RSS at the distance of R and $0.8\,R$ from a BS. R is the cell radius. The centers of the input membership functions for the sets "Very Low" and "Very High" are located at the distance of $2\Delta RSS_{nom}$ dB from RSS_{nom}. The center of the membership function for the fuzzy set "Medium" of the output fuzzy variable $RSS_{threshold}$, $RSS_{threshold(nom)}$, is the RSS at the distance of $1.2\,R$ from a BS. The centers of the membership functions for the sets "Low" and "High" of the output fuzzy variable $RSS_{threshold}$ are located at the distance of $\Delta RSS_{threshold(nom)}$ dB from $RSS_{threshold(nom)}$ where $\Delta RSS_{threshold(nom)}$ is half the difference in RSS at the distance of $1.3\,R$ and $1.2\,R$ from a BS. The centers of the membership functions for the sets "Very Low" and "Very High" of the output fuzzy variable $RSS_{threshold}$ are located at the distance of $2\Delta RSS_{threshold(nom)}$ dB from $RSS_{threshold(nom)}$. The center of the membership function for the fuzzy set "Medium" of the output fuzzy variable $RSS_{hysteresis}$, $RSS_{hysteresis(nom)} = 8$ dB. The centers of the membership functions for the sets "Low" and "High" of the output fuzzy variable $RSS_{hysteresis}$ are located at the distance of $\Delta RSS_{hysteresis(nom)} = 2$ dB from $RSS_{hysteresis(nom)}$. The centers of the membership functions for the sets "Very Low" and "Very High" of the output fuzzy variable $RSS_{hysteresis}$ are located at the distance of $2\Delta RSS_{hysteresis(nom)}$ dB from $RSS_{hysteresis(nom)}$. The spreads of the membership functions are chosen in such a way that the membership value drops to zero at the center of the membership function of the nearest set.

For the secondary FLS, the center of the input membership function for the set "Medium" of the input fuzzy variable *Traffic* is $Traffic_{nom} = 30$. The centers of the input membership functions for the sets "Low" and "High" of the fuzzy variable *Traffic* are located at the distance of $\Delta Traffic_{nom} = 28$ from $Traffic_{nom}$. The center of the input membership function for the set "Medium" of the input fuzzy variable *Velocity* is $Velocity_{nom} = 0$ m/sec. The centers of the input membership functions for the sets "Low" and "High" of the fuzzy variable *Velocity* are located at the distance of $\Delta Velocity_{nom} = 29$ m/sec from $Velocity_{nom}$. The centers of the output membership functions of the fuzzy sets are located equidistant from the centers of the adjacent fuzzy sets. The spreads of the membership functions are chosen in such a way that the membership value drops to zero at the center of the membership function of the nearest set.

Figure 10.5 shows the number of base stations in the active set as a function of distance. The distance is measured from BS 5. Both the conventional and proposed algorithms use more BSs near the cell borders to exploit diversity. The proposed soft handoff algorithm tends to use more BSs in the cell border region, which indicates that a fixed parameter conventional algorithm cannot fully exploit the diversity advantage of soft handoff if not properly tuned. Such tuning uncertainties can be partially compensated by using an adaptive algorithm such as the proposed generic algorithm.

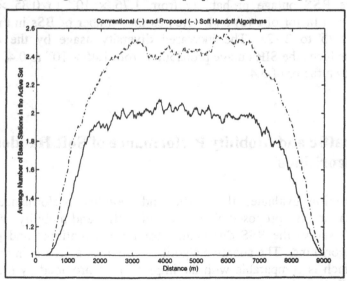

Figure 10.5: Base Stations in the Active Set as a Function of Distance.

The CDF of RSS shown in Figure 10.6 indicates that the increased diversity usage provides a 1.2 dB improvement in the RSS distribution.

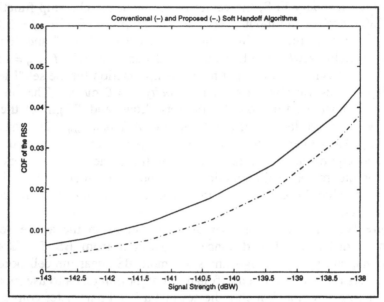

Figure 10.6: Distribution of RSS for the Conventional and Proposed Algorithms.

It is found that the increased diversity usage by the proposed algorithm reduces the RSS outage probability from 1.76×10^{-3} to 0.85×10^{-3} (a reduction by a factor of two) and increases the number of BSs in the Active Set from 1.75 to 2.12. The increased diversity usage by the proposed algorithm reduces the SIR outage probability from 1.45×10^{-3} to 0.43×10^{-3}, a reduction by a factor of 3.4.

10.3.2 Traffic and Mobility Performance of Soft Handoff Algorithms

This section evaluates the traffic and mobility performance of the conventional and the proposed algorithm with traffic and mobility adaptation. Figure 10.7 shows the RSS distribution for the conventional and proposed adaptive algorithms. The adaptive algorithm provides a 1.1 dB improvement in RSS, which is comparable with the improvement provided by the generic soft handoff algorithm. This shows that the adaptation to traffic and mobility does not affect the diversity gain of the generic algorithm significantly since the BSs that meet traffic and mobility requirements still must pass the tests of the conventional soft handoff algorithm with the adaptive handoff parameters supplied by the primary FLS.

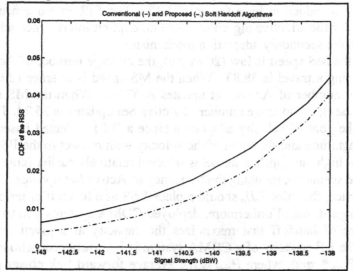

Figure 10.7: RSS Distribution for the Conventional and Proposed Algorithms.

Figure 10.8 shows the traffic distribution for the conventional and proposed adaptive algorithms. The proposed algorithm improves the traffic distribution by two calls, balancing the traffic in the adjacent cells.

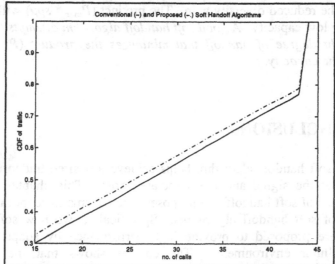

Figure 10.8: Traffic Distribution for the Conventional and Proposed Algorithms.

It is observed that the increased diversity usage by the proposed algorithm reduces the RSS outage probability from 1.76×10^{-3} to 0.8×10^{-3} (a reduction by a factor of two) and increases the number of BSs in the Active Set from 1.75 to 2.12. Also, the SIR outage probability decreases from 1.45×10^{-3} to 1.28×10^{-3}, a reduction by a factor of 1.13. The SIR improvement is not significant for the adaptive algorithm since no explicit interference adaptation is built into the secondary adaptation mechanism.

When the MS speed is low (20 m/sec), the average number of Active Set updates during a travel is 38.83. When the MS speed is average (29 m/sec), the average number of Active Set updates is 37.74. When the MS speed is high (38 m/sec), the average number of Active Set updates is 35.74. This is in line with the goal of velocity adaptation since a BS is selected based on the relative magnitude and direction of the velocity with respect to the BS. When the speed is high, an appropriate BS is selected relatively earlier (compared to a low speed scenario), reducing the frequency of Active Set updates.

In practice, the pilot E_c/I_o should replace RSS as a handoff criterion in the proposed algorithms. Furthermore, deployed CDMA systems aim to provide such degree of handoff that maximizes the capacity at a given quality of service. The RF capacity of a CDMA system is inversely proportional to the product, $(P_{user} * spu)$, where P_{user} is the average forward link power per user that achieves the target frame error rate (FER) and spu is the average number of sectors per user. If the degree of handoff is too high, spu is high, and the required P_{user} will be small because of the diversity gain of soft handoff. However, the product $(P_{user} * spu)$ may be large, leading to low capacity. If the degree of handoff is too low, spu is low, but the required P_{user} will be large because of the reduced diversity gain. The product $(P_{user} * spu)$ will again be large, giving low capacity. *A good soft handoff algorithm attempts to provide just the right degree of handoff that minimizes the product $(P_{user} * spu)$, increasing the capacity.*

10.4   CONCLUSION

A good soft handoff algorithm helps achieve a desired tradeoff between the quality of the signal and the associated cost. This chapter considers several aspects of soft handoff and proposes mechanisms to adapt the handoff parameters of soft handoff algorithms. Specifically, two new soft handoff algorithms are proposed to provide high performance by adapting to the dynamic cellular environment. The chapter shows that the proposed algorithms address critical system design issues related to soft handoff. In particular, the proposed algorithm enhances the degree of diversity exploitation if the associated cost (e.g., traffic in a cell) is not high.

Chapter 11

RADIO RESOURCE MANAGEMENT AND EMERGING CELLULAR SYSTEMS

Initially deployed cellular systems were analog and constituted the first generation cellular systems. Growth of voice subscribers led to high capacity digital systems called second generation (2G) systems. Examples of 2G systems are narrowband CDMA-based IS-95 systems and TDMA-based GSM systems. Demand for increased voice capacity and expected growth of data subscribers are now driving the development of emerging third generation (3G) and third-and-a-half generation (3.5G) systems. Two dominant 3G cellular standards are Interim Standard-2000 (IS-2000) and UMTS (Universal Mobile Telephone System). Radio interface related features of these standards along with RRM aspects of these systems are the focus of this chapter. A standard to specifically address the wireless data market has been developed and is called 1xEvolution-Data Only (1xEV-DO). The 1xEV-DO standard (also referred to as IS-856) is considered a 3.5G system and represents the first phase in the evolution of IS-2000. The second phase of the evolution of IS-2000 is 1xEvolution-Data and Voice (1xEV-DV). Prominent features and distinct RRM requirements of the 1xEV-DO are highlighted. Methodology for fuzzy logic based and neural network based handoff algorithms developed in this book is viewed from the perspective of RRM.

11.1 INTRODUCTION TO EMERGING SYSTEMS

Widespread acceptance of the Internet and Internet-related services has led to expectations for such services in the wireless domain as well. The transition from 2G systems to 3G systems was fueled largely by required voice capacity increases and emerging data services. Third generation systems such as IS-2000 systems in the U.S. [115] and UMTS systems in Europe [120] will be deployed from Year 2001 onward. Two groups instrumental in the standardization of UMTS and IS-2000 are Third Generation Partnership Project (3GPP) [http://www.3gpp.org/] and Third Generation Partnership Project 2 (3GPP2) [http://www.3gpp2.org/]. The

3GPP specified UMTS Terrestrial Radio Access (UTRA) that is based on an evolved GSM core network and the CDMA technology. The 3GPP2 specified IS-2000 that is based on evolution of IS-95. In June 1999, several major operators in the Operator Harmonization Group (OHG) proposed a harmonized Global 3G (G3G) concept that was later accepted by 3GPP and 3GPP2. The harmonized G3G concept aims for a single global standard with three modes of operation, CDMA-Direct Spread (DS) mode based on UTRA-FDD, CDMA-Multi-Carrier (MC) mode based on IS-2000, and CDMA-TDD mode based on UTRA-TDD [122]. The major goal of 3G and 3.5G technologies is enhanced data performance. Work has already commenced toward the development of next generation technologies, i.e., 3.5G and beyond. The 1xEV-DO standard, an example of 3.5G technology, is based on Qualcomm's high data rate (HDR) proposal [121]. Further enhancements to IS-2000 systems, more specifically, 1.25 MHz systems, are also currently being investigated to create a 1xEV-DV standard. This chapter discusses two 3G technologies, IS2000 and UMTS, and one 3.5G technology, 1xEv-DO. RRM for these systems is also discussed.

Section 11.2 provides an introduction to IS-2000. Generic radio features of IS-2000 are summarized in Section 11.3, while RRM for IS-2000 is discussed in Section 11.4. The UMTS is introduced in Section 11.5. Section 11.6 mentions features of UMTS, and Section 11.7 considers RRM for UMTS. The 1xEV-DO is the focus of Section 11.8, and RRM for 1xEV-DO is discussed in Section 11.9. The neural and fuzzy handoff techniques are generalized from the RRM perspective in Section 11.10. Finally, Section 11.11 summarizes the chapter.

11.2   INTRODUCTION TO IS-2000

To evolve the Direct Sequence- CDMA (DS-CDMA) based IS-95 standard, the IS-2000 standard has been developed. The IS-2000 standard is also referred to as cdma2000. The overall system architecture for an IS-2000 system is shown in Figure 11.1. The MS communicates with one or more BSs. A BSC controls multiple BSs. Functions requiring input from multiple BSs, e.g., soft handoff and power control, are implemented at the BSC. The MSC oversees the operation of the BSCs and interfaces the wireless system to the wireline system such as the Public Switched Telephone Network (PSTN) for telephone services and the PSDN (Packet Switched Data Network) for the Internet services.

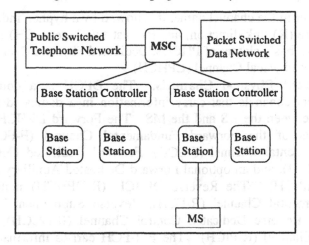

Figure 11.1: System Architecture for IS-2000.

Radio interface related processing is performed on physical channels (PHCHs) in the IS-2000 system shown in Figure 11.2. The IS-2000 physical layer defines several PHCHs that are transmitted over the air.

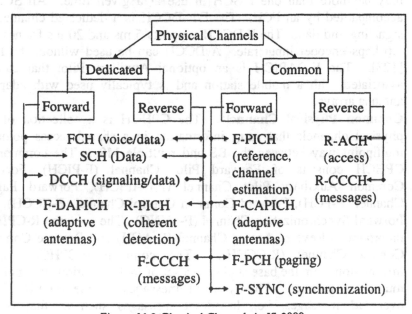

Figure 11.2: Physical Channels in IS-2000.

The first letter in the channel name, if followed by a hyphen indicates forward (F) or reverse (R) link direction, and the last two letters (CH) mean channel. There are two basic classes of PHCHs, Dedicated Physical Channel (DPHCH) and Common Physical Channel (CPHCH).

- **Dedicated Physical Channels**. The DPCH is a collection of all physical channels that carry information in a dedicated point-to-point way between the BS and the MS. The Forward DPHCH (F-DPHCH) consists of the Forward Fundamental Channel (F-FCH), Forward Supplemental Channel (F-SCH), Forward Dedicated Control Channel (F-DCCH), and an optional Forward Dedicated Auxiliary Pilot Channel (F-DAPICH). The Reverse DPHCH (R-DPHCH) includes Reverse Fundamental Channel (R-FCH), Reverse Supplemental Channel (R-SCH), Reverse Dedicated Control Channel (R-DCCH), and Reverse Pilot Channel (R-PICH). The F/R-FCH carries information (voice or data), is transmitted at variable rates, and requires rate detection at the receiver. The F/R-SCH is a high rate channel. The F-SCH can be operated in two modes. In the first mode (applicable only for data rates up to 14.4 kbps), blind rate detection is used. In the second mode of operation, rate information is explicitly provided to the receiver. There may be more than one F-SCH in use at a given time. An SCH is accompanied by an FCH. The F/R-DCCH is a dedicated channel for signaling and data. The F-DCCH supports 5 ms and 20 ms frames at a 9.6 kbps encoder input rate. A DCCH can be used without the FCH [123]. The F-DAPICH is an optional Auxiliary Pilot that can be associated with a mobile station and is typically used with adaptive antenna arrays.

- **Common Physical Channels**. The CPHCH is a collection of the physical channels that carry information in a shared-access point-to-multipoint way between the BS and multiple MSs. The common F-CPHCH consists of Forward Pilot Channel (F-PICH), Forward Common Auxiliary Pilot Channel (F-CAPICH), Forward Paging Channel (F-PCH), Forward Common Control Channel (F-CCCH), and Forward Synchronization Channel (F-SYNC). The common R-CPHCH incorporates Reverse Access Channel (R-ACH) and Reverse Common Control Channel (R-CCCH). The common F-CPHCH carries information from the base station to a set of mobile stations in a point to multipoint manner. The types of messages on the F-CPHCH are overhead messages, i.e., broadcast messages such as the System Parameters Message, and directed messages (e.g., a paging message to a single mobile station). The R-CPHCH carries information from multiple mobile stations to the base station on a contention basis. There can be multiple F-PCHs per base station. The F-PCH can transmit at 9.6 kbps or 4.8 kbps. The R-ACH and R-CCCH are reverse common channels used for communication of Layer 3 and MAC messages from the mobile station to the base station. The mobile stations transmit R-ACH and R-CCCH without explicit authorization by the base station. The F/R-CCCH is used for communication of Layer 3 and MAC messages between the base station and the mobile station. The F-CAPICH is a forward link channel associated with an auxiliary pilot. The F-PICH is used for estimation of channel gain and phase, detection of multipaths, and cell acquisition and handoff, while the R-PICH is used for initial acquisition, time tracking, rake-receiver coherent reference recovery, and power-control measurements. The R-PICH consists of a fixed reference value and time-multiplexed power control information (referred to as power control sub-channel). The F-SYNC is used by MSs to acquire initial time synchronization.

11.3 GENERIC FEATURES OF IS-2000

Some prominent features of an IS-2000 system are mentioned below.

- **Configurations.** The system can be a multi-carrier configuration, e.g., three carriers with 1.25 MHz bandwidths to cover a 5 MHz spectrum, or a direct spread configuration, e.g., a single carrier with a 3.75 MHz bandwidth to cover a 5 MHz spectrum. The forward link supports chip rates of N x 1.2288 Mcps (N = 1, 3, 6, 9, and 12). For chip rates corresponding to $N > 1$, chip rates can be obtained using multi-carrier or direct spread configuration. In the multi-carrier configuration, modulation symbols are de-multiplexed onto N separate 1.25 MHz carriers. The direct spread configuration transmits modulation symbols on a single carrier that is spread with a chip rate of N x 1.2288 Mcps. When the IS-2000 system uses a 1.25 MHz bandwidth (similar to the IS-95 system), it is referred to as a 1x Radio Transmission Technology (1xRTT) system. Initial commercial systems will be based on the multi-carrier configuration with $N = 1$ (i.e., 1xRTT) and will occupy 1.25 MHz spectrum in each link. The FDD is used to support simultaneous forward and reverse links.
- **Auxiliary Pilot.** When a system deploys antenna arrays and antenna transmit diversity, a separate pilot called auxiliary *pilot* is required for channel estimation and phase tracking. Auxiliary pilots use orthogonal Walsh codes. To alleviate potential Walsh code exhaustion, a longer Walsh sequence may be used for auxiliary pilots.
- **Data Rates.** The 1.25 MHz system supports data rates up to 307.2 kbps, and the 5 MHz system supports data rates up to 1.0368 Mbps for a single

channel on the forward link. The reverse link supports 230.4 kbps for the 1.25 MHz system and 1.0368 Mbps for the 5 MHz system for a single channel. Multiple channels can be used for data transmission, i.e., the combination of a fundamental channel and a supplemental channel or the combination of multiple supplemental channels.

- **MAC.** Advanced MAC is implemented for efficient data applications. The MAC provides mechanisms, e.g., timers, to manage physical layer resources.
- **Radio Interface.** The radio interface of the IS-2000 system is an enhanced version of the radio interface of the IS-95 system. Salient physical layer features include coherent (pilot-based) reverse link, fast power control in both forward and reverse links, and availability of turbo coding in addition to convolutional coding.
- **Power Control.** Both forward and reverse links can implement 800 Hz power control for the FCH. When both FCH and SCH are transmitted in the forward link, 800 Hz power control can be shared between the FCH and the SCH, e.g., 400 Hz each for FCH and SCH. A general block diagram of the power control operation is illustrated in Figure 11.3.

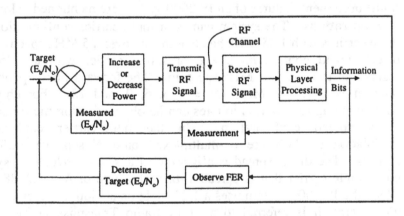

Figure 11.3: Power Control Operation in IS-2000.

In the forward link, the MS compares the received E_b/N_o with a target E_b/N_o, and asks the BS to increase or decrease power by a specific step size (a configurable parameter that can be 1.0 dB, for example) every power control group, i.e., every 1.25 ms, as part of inner loop power control. An outer loop changes the target E_b/N_o every frame, i.e., every 20 ms, by a fixed configurable amount such that a desired long-term FER is met. The SCH power can be controlled independent of the FCH power for the forward link. Such independent power control of an SCH is efficient since important power control parameters such as FER targets and channel coding are usually different for the FCH and the SCH. The

SCH power can also be adjusted relative to the FCH power. The reverse link power control is similar to the forward link power control. The base station asks the MS to increase or decrease by a predetermined step size such that a target E_b/N_o is met. An outer loop adjusts the setting of the E_b/N_o every frame based on the quality of the frame. In the reverse link, both the FCH and SCH power control is tied together. In other words, the FCH is power controlled at 800 Hz, and the SCH power is adjusted based on a configurable ratio that relates FCH power to the SCH power.

- **User Distinction.** Users are distinguished by Walsh codes on the forward link. There are user-specific long codes on the reverse link. The IS-2000 system uses short PN codes for scrambling on both the forward link and reverse link. All BSs use the same PN code with different PN offsets.

- **Short Data Bursts.** Short Data Bursts (SDBs) are messages or data consisting of a small number of frames that are transmitted on a common channel when a dedicated traffic channel is not.

- **Modulation.** Each forward link physical channel is modulated by a Walsh code, and QPSK modulation is used. The Walsh code length varies for different information bit rates. When the performance is limited by the Walsh codes, quasi-orthogonal functions can be used as additional Walsh codes. On the reverse link, BPSK modulation is used.

- **Physical Layer Processing for the Forward Link.** The basic steps involved in the processing of information bits for the FCH and the SCH are addition of reserved bits, addition of frame quality indicator, addition of encoder tail bits, convolutional or turbo coding, symbol repetition, symbol puncturing, block interleaving, user-specific scrambling, power control puncturing (for FCH and DCCH), QPSK modulation, Walsh code spreading, and BS-specific PN code masking. For the forward link of the multi-carrier System, the user data is scrambled using the user-specific code, and the scrambled data is de-multiplexed onto N carriers (N=3, 6, 9, or 12). Power control bits may be punctured on the forward link channels at the rate of 800 Hz. The signal on each carrier is orthogonally spread by an appropriate Walsh code function. Different Walsh codes can be used on different carriers. In case of the forward link for the direct-spread system, the interleaver output data is scrambled using the user-specific long code. The scrambled data undergoes in-phase (I) and quadrature (Q) mapping, channel gain multiplication, power control puncturing, and Walsh spreading.

- **Physical Layer Processing for the Reverse Link.** The basic steps involved in the processing of information bits for the FCH and the SCH are addition of reserved bits, addition of frame quality indicator, addition of encoder tail bits, convolutional or turbo coding, symbol repetition, symbol puncturing, block interleaving, channel-specific orthogonalization, I and Q mapping, BPSK modulation, user-specific long code masking, and spreading. The spread pilot and R-DCCH are mapped into the I data channel, while the spread R-FCH and R-SCH are mapped into the Q data channel. The I and Q data channels are spread using a complex-multiply PN spreading approach. The levels of the fundamental, supplemental, and dedicated control channels are adjusted relative to the reverse pilot channel.

- **Transmit Diversity.** There is no mandatory transmit diversity technique. Note that a multi-carrier system inherently provides frequency diversity because different carrier frequencies are transmitted on different antennas. Examples of optional transmit diversity techniques include Orthogonal Transmit Diversity (OTD) and Phase Sweeping Transmit Diversity (PSTD) [124]. Some transmit diversity techniques, e.g., OTD, do require a change in the standard; however, techniques such as PSTD do not require any change in the standard.

- **Frame Length.** The IS-2000 system supports 5 ms and 20 ms frames for control information on the fundamental and dedicated control channels and 20 ms frames for other types of data including voice. The frame length of 5 ms is used for only R-FCH. Interleaving and sequence repetition are performed over the frame interval.

- **Handoff.** The IS-2000 system supports soft handoff, as does the IS-95 system, on both the forward and reverse link.

- **IS-95 and IS-2000**. The IS-2000 system provides full backward compatibility with IS-95. Hence, an IS-95 terminal can be served by a IS-2000 system.

11.4 RADIO RESOURCE MANAGEMENT IN IS-2000

In the forward link, an FCH carries control information and data. A user with an FCH is said to be in an Active state. Since the full rate FCH is a low rate channel, i.e., 9.6 kbps or 14.4 kbps, an SCH is used to transmit a large amount of data. A high rate SCH consumes more power and requires more Walsh codes, and, hence, only a limited number of high rate SCHs can be supported simultaneously. The FCH and the SCH are separately coded and interleaved and have different transmit power levels and FER targets. A higher FER channel requires less power than a lower FER channel, leading to power savings. A detailed analysis of power savings can be found in [116]. Note that a data service, in general, can tolerate higher FER than a voice service due to upper layer retransmissions. Each active user receives data on a dedicated FCH, while SCHs are time-shared by active users. Thus, the combination of the FCH and the SCH is a common approach for packet data transmission in the forward link. Another approach for packet data transmission is to use the combination of a DCCH and an SCH. The advantage of the second approach is that the power is saved by discontinuous transmission in the absence of data, but this approach complicates power control [123]. Note that the FCH transmits eighth rate frames in the absence of data in the first approach. If there are many users receiving eighth rate frames from the BS, a significant amount of precious forward link power is wasted. To utilize power more efficiently in the first approach, a mechanism that controls use of the FCH can be implemented. The IS-2000 standard provides a MAC timer, called T_{active}, that can be used by such a mechanism to perform a transition from an Active state to a Dormant state. In the Dormant state, the user does not have an FCH. Thus, the FCH is released in the absence of data. If T_{active} is too short and if data arrives soon after the Active state to Dormant state transition, a Dormant to Active state transition may occur, and a significant delay is experienced by the user. The rate of the SCH is controlled by the network to achieve a balance between the service requirements and the need for sharing the resources. For example, a low rate SCH can be used to save some power for other users.

In the reverse link, one way to transmit packets is through the combination of the FCH and the SCH (as in case of the forward link). The Active and Dormant state transitions are handled similarly to the forward link case. The reverse link SCH rate may or may not be controlled by the network. If the reverse link becomes a capacity bottleneck, reverse link loading becomes critical, and downgrading of the R-SCH rate or rejection of the SCH transmission request is likely to occur. If the deployed IS-2000 systems are found to be forward link limited (similar to existing IS-95 systems), extensive management of resources for the reverse link is not required. In general, the enhanced reverse link of IS-2000, e.g., coherent detection compared to non-coherent detection in IS-95, and the asymmetric nature of prominent data services, e.g., web browsing and e-mail reading, are less likely to make the reverse link a capacity bottleneck.

The packet data transmission process described above indicates that the forward link RRM in a CDMA system should at the very least control resources such as power and Walsh codes. Since the forward link is the limiting link in a typical CDMA system, only forward link RRM is discussed here. Management of power is very crucial for IS-2000 systems because power is the performance bottleneck for the forward link. Two major components of the IS-2000 RRM are *call admission control* and *SCH management*. Figure 11.4 depicts the overall RRM for IS-2000. When a new call arrives, it is admitted if there is sufficient power and an available Walsh code. If the data call cannot be admitted, it is possible to queue the call for some time period. Since delay cannot be tolerated for a voice call, the voice call is blocked in the absence of required resources.

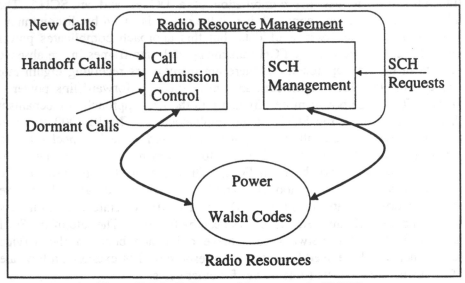

Figure 11.4: Radio Resource Management for IS-2000.

Each newly admitted call is assigned an FCH. If there is no data activity for some time period (equal to T_{active}), the user enters the Dormant state. In the Dormant state, the user does not have an FCH. Data activity triggers the Dormant to Active state transition. The system performs the transition if an FCH can be assigned. Otherwise, such a request can be queued until resources become available. Thus, the call admission control mechanism deals with new calls as well as calls requesting Dormant to Active transitions. In practice, the call admission control process also considers the impact of handoffs. After the handoff decision is made by the network, the call admission control mechanism assumes the charge of resource allocation for the handoff call. A handoff call is usually prioritized over a new call because a dropped call is considered more annoying to the user than a blocked call. In a CDMA system, handoff prioritization can be achieved by reserving some power that cannot be accessed by new calls.

The SCH management is required for only Active users, i.e., users with FCHs. Since a low rate FCH cannot handle a large amount of data, a high rate SCH aids transmission of data as explained earlier. The rate of the SCH and the time duration for which the user has an exclusive access to the SCH (often called burst duration) depend on the service requirements, data in the buffer, data arrival rate, available power, and available Walsh codes. The IS-2000 standard provides a set of values for the burst duration. The choice of the rate and the burst duration represents a balance between two competing needs: to serve the current user with the maximum possible rate for a long period and to share the resources among all users with some degree of fairness.

Since the frequency reuse factor is one in a CDMA system, the same carrier frequency is used in all cells (or sectors). For example, the same 1.25 MHz wide spectrum, e.g., around a carrier frequency of 900 MHz or 1900 MHz, is used for the forward link in all the cells (or sectors if there are multiple sectors per cell) in the system. The forward link is the capacity bottleneck in a CDMA system because of the limited power and the nature of forward link interference. If the subscriber demand cannot be met using the single carrier frequency, another carrier frequency, i.e., additional 1.25 MHz spectrum in the forward link, is often deployed, creating a multi-carrier cellular system. For a multi-carrier system, the call admission control mechanism can assign any carrier frequency to the newly arrived call. The existence of multiple carriers provides an opportunity for traffic balancing (or load balancing) between the carriers. A technique called multi-carrier traffic allocation (MCTA) can be employed to balance power usage for different carriers, increasing the capacity of the system, compared to the system that randomly allocates a carrier to the call [125]. The MCTA assigns to the new call a carrier frequency that has lowest amount of total power in use among all carrier frequencies. For example, assume that there are ten users in a sector on each carrier frequency and that the total power in use is 70% on one frequency and 95% on the other frequency. If a new call is directed to the second carrier frequency, it may be blocked because of insufficient power on the second frequency. However, the call may be successfully admitted into the system if it is directed to the first carrier frequency, because there is a sufficient amount of power available to serve the user. Recall that the availability of the forward link power is the most critical component in a call admission algorithm for a CDMA system. In deployed IS-95 systems, calls are blocked primarily due to insufficient forward link power. Thus, at a given call blocking probability, an MCTA system could support more users compared to a non-MCTA system that randomly assigns the frequency to the call. In summary, the MCTA increases the system capacity by exploiting variations in power requirements of users. Even when there are the same numbers of users on the carrier frequencies, significant differences in the total amounts of power transmitted by the BSs on the carrier frequencies exist because of the influence of the MS location and the propagation channel conditions on the power required by a user. The MCTA technique is a good example of traffic (or load) balancing in a CDMA system.

11.5 INTRODUCTION TO UMTS

The goal of UMTS is to provide global mobile multimedia communications. To complement a terrestrial system, a satellite component is also planned, called Satellite UMTS or S-UMTS, that will provide global coverage and potentially traffic balancing [126]. The UMTS architecture consists of the Core Network (CN), UTRA Network (UTRAN), and User Equipment (UE) as shown in Figure 11.5.

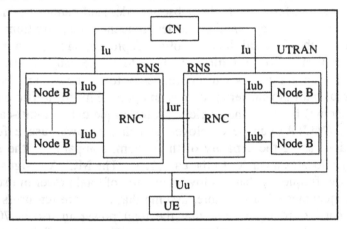

Figure 11.5: UMTS Architecture.

In a UMTS system, the UE refers to the mobile terminal or user terminal. The UE contains one or several UMTS Subscriber Identity Modules (SIMs). The SIM identifies users and can perform functions such as storage of dialing numbers. The CN handles tasks such as switching, subscriber data base management, billing, and call management (e.g., roaming) and provides an interface between the cellular system and the wire-line networks such as the PSTN, which provides telephone services and the PSDN, which provides Internet access. The UTRAN consists of Node Bs and Radio Network Controllers (RNCs). A Node B is a logical node responsible for radio transmission/reception in one or more cells to/from the UE. An RNC controls the use of radio resources. Thus, a Node B can be viewed as a BS or a group of BSs, and an RNC can be considered as a base station controller that controls several BSs. The communications link from the Node B to the UE is called downlink, while the communications link from the UE to the Node B is referred to as uplink. A Radio Network Subsystem (RNS) is the access part of a UMTS network that manages resources to establish connections between a UE and the UTRAN. An RNS contains one RNC and one or more Node Bs and is responsible for the resources and transmission/reception in a set of cells. A UMTS system defines several interfaces, e.g., Iu, Iub, Uu, and Iur, as depicted in Figure 11.5. Iu is the interconnection point between the RNS and the CN. Iub is the interface between the RNC and Node B, while Uu is the interface between UTRAN and the UE. The interface between RNSs is called Iur.

Radio interface in the UMTS consists of Layers 1, 2, and 3. Layer 1, the Physical Layer, transports encoded and spread chips over the air. Layer 2 consists of two sub-layers, the MAC layer and Radio Link Control (RLC) layer. The MAC layer maps logical channels into transport channels (discussed later), and the RLC provides segmentation/re-assembly, ciphering, and reliability over the radio link. The Radio Resource Control (RRC) controls how resources of Layer 1, e.g., physical channels, are used. The multiple access scheme for the radio interface is DS-CDMA with a 5 MHz bandwidth. The UMTS supports two modes, FDD-CDMA and TDD-CDMA. Layers 2 and 3 of the radio interface are common for different physical layers corresponding to these two different modes. The FDD-CDMA requires paired bands, while the TDD-CDMA needs just a single band. The UTRA FDD and TDD share the same basic system parameters such as 3.84 Mcps chip rate, QPSK modulation (in forward link), 5 MHz signal bandwidth, root raised cosine pulse shaping, 10 ms frame length, and 15 time slots per frame. A service provider can choose the mode based on spectrum availability, coverage requirements, data requirements, and mobility needs. In general, the FDD mode is suitable for large, i.e., macro and micro, cells, high degrees of mobility, data rates up to 384 kbps, and symmetric data rates on the forward link and the reverse link. The TDD mode is suitable for small, i.e., micro and pico, cells, low degrees of mobility, data rates up to 2 Mbps, and asymmetric data rates.

Figure 11.6 illustrates two parts of the radio interface related processing. The radio processing is largely performed on channels called Transport Channels (TrCHs) that are different from physical channels. Some radio processing is performed on the physical channels.

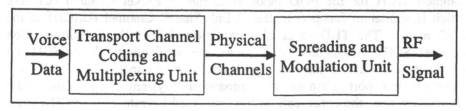

Figure 11.6: Radio Interface Processing in UMTS.

The TrCHs are suitably mapped onto physical channels. A transport channel is defined by the data characteristics and the method of data transmission over the air. Transport channels convey control plane or user plane data between the UE and the RNC [122]. The transport channels are summarized in Figure 11.7. There are two types of transport channels, dedicated channels and common channels.

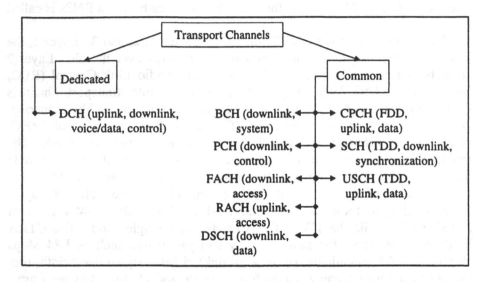

Figure 11.7: Transport Channels in UMTS.

The FDD and TDD modes define one type of dedicated transport channel, the Dedicated Channel (DCH). The DCH is an uplink or downlink channel. The DCH is transmitted over the entire cell or part of the cell when beam-forming is used. The DCH involves the potential for fast data rate changes (every 10 ms), fast power control, and inherent addressing of UEs. Six common TrCHs are defined in the FDD mode. Five of such TrCHs are the Broadcast Channel (BCH), Paging Channel (PCH), Forward Access Channel (FACH), Random Access Channel (RACH), and Downlink Shared Channel (DSCH). These TrCHs are also defined in the TDD mode. The remaining common TrCH for the FDD mode is Common Packet Channel (CPCH), which is similar in function to the Uplink Shared Channel (USCH) in the TDD mode. The TDD mode also explicitly defines the Synchronization Channel (SCH). Although the FDD mode does not explicitly define SCH as a TrCH, it does have a physical synchronization channel. The BCH is a downlink transport channel that broadcasts system- and cell-specific information over the entire cell with a low fixed bit rate. The PCH carries control information to the UE when the system does not know the location cell of the UE. The FACH is a downlink transport channel transmitted over the whole cell or part of the cell for beamforming applications and uses slow power control. The RACH carries control information from the UE and may also carry short user packets. The DSCH is a downlink transport channel shared by several UEs and associated with a DCH. The CPCH (or the USCH) is a common uplink transport channel that can be used by UEs for power-controlled transmission of data traffic. The SCH is used for synchronization.

As mentioned earlier, much of the radio processing is performed on the TrCHs. The basic steps involved in the processing, specifically channel coding and multiplexing, of TrCHs are very similar for FDD and TDD modes. Data is in the form of transport block (TrBk) sets and is sent to the coding/multiplexing unit once every transmission time interval (TTI). The TTI is 10 ms, 20 ms, 40 ms, or 80 ms based on the type of the TrCH and type of service. For example, coded voice may use 20 ms TTI for low end-to-end delay, while packet data may use 80 ms TTI to exploit superior performance of turbo coding for large data block sizes. One radio frame from each TrCH is sent to the TrCH multiplexing every 10 ms. These radio frames are serially multiplexed to create a Coded Composite Transport Channel (CCTrCH). A CCTrCH is mapped to one or more physical channels.

The steps followed in the *coding/multiplexing unit for the FDD uplink and TDD uplink and downlink* for each transport channel are addition of CRC to the transport block, transport block concatenation and code block segmentation, channel coding, radio frame equalization, first stage interleaving, radio frame segmentation, and rate matching. Multiplexing of such processed TrCHs is performed to create a CCTrCH. The processing of the CCTrCH is different for FDD uplink and TDD uplink and downlink. In the FDD uplink and TDD uplink and downlink, the three major steps executed are physical channel segmentation, second stage interleaving, and physical channel mapping. In the TDD mode, the second stage interleaving can be applied to all data bits in a frame or to bits in a time slot. The selection of the frame-based or the time-slot-based interleaving is controlled by higher layer. Furthermore, the physical channel mapping for the TDD mode is done according to the type of the second stage interleaving.

The steps followed in the *coding/multiplexing unit on the FDD downlink* for each transport channel are addition of CRC to the transport block, transport block concatenation and code block segmentation, channel coding, rate matching, insertion of discontinuous transmission (DTX) indication bits (fixed positions), first stage interleaving, and radio frame segmentation. Multiplexing of transport channels creates a CCTrCH. The CCTrCH undergoes insertion of DTX indication bits, physical channel segmentation, second stage interleaving, and physical channel mapping.

The end result of the TrCH processing is a set of physical channels. Further radio related processing is done on the physical channels according to the mode, i.e., FDD or TDD, as explained later. A physical channel consists of frames and time slots. A radio frame is of 10 ms duration and consists of 15 time slots. The configurations of the radio frames and time slots and information contents during a given time slot depend on symbol rate and type of the physical channel and the mode. In the UTRA-TDD mode, the maximum number of basic physical channels per frame for the downlink is given by the product of the maximum number of time slots, i.e., 15, and the maximum number of CDMA codes per time slot, i.e., 16. In the case of UTRA-FDD mode, the maximum number of basic physical channels per frame for the downlink is the spreading factor of 256, i.e., a maximum of 256 distinct Orthogonal Variable Spreading Factor codes. Some slot formats allow SF of 512, but the user data rate is very low, e.g., 3 or 6 kbps.

11.6   GENERIC FEATURES OF UMTS

Prominent features of a UMTS system are outlined below.

- **Infrastructure.** The UTRAN connects to the enhanced GSM Network and Switching Subsystem (NSS) to serve UMTS subscribers along with GSM subscribers. The GSM radio subsystem and UTRAN are two access systems that use the same network infrastructure.

- **Subscriber Support in UMTS and GSM.** The SIMs in UMTS and GSM are exchangeable. A GSM SIM can be inserted into a UMTS terminal to access a UMTS system. A UMTS SIM in a dual mode UMTS/GSM terminal can be served in a 2G GSM-only environment [126].

- **Mobility and Data Rate Support.** The UTRA supports at least 144 kbps for mobility at speeds up to 500 km/hr, 384 kbps for limited mobility up to 120 km/hr in macro- and micro-cellular environments, and 2.048 Mbps for low-mobility applications up to 10 km/hr, in micro- and pico-cellular outdoor and indoor environments.

- **Handovers.** Handovers are allowed between GSM, UTRA-TDD, and UTRA-FDD.

- **Physical Layer Attributes.** Both convolutional and turbo codes are used for channel coding. QPSK modulation is implemented on the forward link, and BPSK modulation is used on the reverse link.

- **Orthogonalization and Source Distinction.** To separate channels from the same source, e.g., the Node B or the UE, channelization codes are used. To distinguish two cells, Gold codes with 10 ms periods are used in the FDD mode and scrambling codes are used in the TDD mode. To differentiate two UEs, Gold codes with 10 ms periods or S(2) codes with 256 chip periods are used for the FDD mode, while codes and midambles are used for the TDD mode. Orthogonal, e.g., Walsh-Hadamard, codes provide channelization.

- **Discontinuous Radio Transmission.** When voice, data, or control information is absent, the mobile discontinues transmission to reduce interference between UEs in the FDD mode. The TDD mode utilizes time-slots, leading to inherent discontinuous transmission. In the TDD mode, this mechanism is applied in both uplink and downlink. If no data is present in a resource unit, no transmission takes place.
- **Transport Format Detection.** Detection of the transport format can be blind or through explicit transmission of the Transport Format Combination Indicator (TFCI). The TFCI provides the receiver with the information on the parameters of different TrCHs multiplexed on the physical channels and corresponds to the data transmitted in the same frame.

11.6.1 UTRA-FDD Features

Prominent features of UTRA-FDD are briefly discussed here.
- **Dedicated Uplink Physical Channels.** There are two types of uplink dedicated physical channels, the uplink Dedicated Physical Data Channel (uplink DPDCH) and the uplink Dedicated Physical Control Channel (uplink DPCCH). Figure 11.8 shows the physical channels defined in UTRA-FDD.

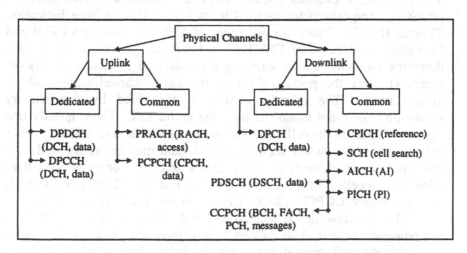

Figure 11.8: Physical Channels in UTRA-FDD.

The DPDCH and the DPCCH are I and Q multiplexed within each radio frame. The uplink DPDCH carries data generated at the DCH. The uplink DPCCH carries control information generated at Layer 1. This control information includes known pilot bits (for coherent detection), transmit power control (TPC) commands, feedback information (FBI), and an optional TFCI. The UTRAN dictates to the UE whether or not to send the TFCI. All UEs must support the TFCI. The time slot has 2560 chips and $10*2^k$ bits (where k is from 0 to 6). The SF of a physical channel is given by $256/2^k$. Hence, the SF of the DPDCH ranges from 256 to 4.

- **Common Uplink Physical Channels.** These include Physical Random Access Channel (PRACH) and Physical Common Packet Channel (PCPCH). The PRACH carries the RACH and utilizes a slotted ALOHA approach with fast acquisition indication. The PCPCH carries the CPCH.

- **Dedicated Downlink Physical Channels.** There is only one type of downlink dedicated physical channel, the downlink Dedicated Physical Channel (downlink DPCH). Within a given downlink DPCH, the DCH (the transport channel) is transmitted in time-multiplex with control information generated at Layer 1 (known pilot bits, TPC commands, and an optional TFCI). In other words, the downlink DPCH is a time multiplex of a downlink DPDCH and a downlink DPCCH.

- **Common Downlink Physical Channels.** They include Common Pilot Channels (CPICHs), Common Control Physical Channels (CCPCHs), Synchronization Channel (SCH), Physical Downlink Shared Channel (PDSCH), Acquisition Indication Channel (AICH), and Page Indication Channel (PICH). There are two types of CPICHs, Primary CPICH and Secondary CPICH. The CPICH is a fixed rate (30 kbps, SF = 256) downlink physical channel carrying a pre-defined sequence of bits (or symbols). For the primary CPICH, the same channelization code is always used. The primary CPICH is scrambled by the primary scrambling code and broadcast over the entire cell. There is only one Primary CPICH in the cell. The Secondary CPICH is scrambled by either the primary or secondary scrambling code and may be transmitted over the entire or part of the cell. There can be zero, one, or more Secondary CPICHs in a cell. There are two CCPCHs, Primary CCPCH (P-CCPCH) and Secondary CCPCH (S-CCPCH). The main difference between a CCPCH and a downlink dedicated physical channel is that a CCPCH is not power-controlled. The P-CCPCH is a fixed-rate (30 kbps, SF = 256) downlink physical channel that caries the BCH. The P-CCPCH consists of only data and does not include TPC, TFCI, and pilot bits. Furthermore, the P-CCPCH is switched off during the first 256 chips of each time slot, such that the Primary SCH and Secondary SCH can be transmitted during this period. The S-CCPCH carries the FACH and PCH and may or may not include TFCI. The FACH and the PCH can be mapped to the same or different S-CCPCHs. The S-CCPCH can support variable rates with the help of the TFCI. The P-CCPCH is continuously transmitted over the entire cell, while the S-CCPCH is transmitted only when data is available. The SCH is a downlink channel for cell search. The SCH consists of the Primary SCH and Secondary SCH, which are transmitted in parallel. The Primary Synchronization Code (PSC) transmitted in the Primary SCH is the same for every cell in the system. The Secondary Synchronization Code (SSC) indicates which of the code groups the cell's downlink scrambling code belongs to. The PDSCH carries the DSCH and is shared by users based on code multiplexing. The PDSCH is always associated with a downlink DPCH. The spreading factor of the PDSCH may vary from frame to frame. A DSCH may be mapped to multiple parallel PDSCHs. The relevant Layer 1 control information is transmitted on the DPCCH part of the associated DPCH. The AICH is a common downlink physical channel that carries Acquisition Indicators (AIs). The AI corresponds to the signature on the PRACH or PCPCH. The PICH is a common downlink fixed rate physical channel that carries Page Indicators (PIs). The PICH is associated with an S-CCPCH.

- **Diversity Techniques.** In the downlink, Space Time Transmit Diversity (STTD) and Time Switched Transmit Diversity (TSTD) are optional in the UTRAN but mandatory at the UE. Transmit diversity in the open loop mode can be applied to downlink physical channels; however, the closed loop option is available for DPCH and PDSCH (associated with DPCH) only.

- **Timing of Physical Channels.** The P-CCPCH provides time reference for other physical channels.

- **Spreading and Modulation for the Uplink Physical Channels.** Once the coding/multiplexing unit for TrCH generates physical channels, spreading and modulation of the physical channels occur. Spreading consists of channelization and scrambling. During channelization, a data symbol is transformed into chips, increasing the signal bandwidth. The number of chips per data symbol is referred to as the SF. Data symbols on I and Q branches are independently multiplied by Orthogonal Variable Spreading Factor (OVSF) codes. Channelization codes are OVSF codes that provide orthogonality between a user's different physical channels. These codes are identical to Walsh codes generated using Hadamard matrices. During the scrambling process, a scrambling code is applied to the spread signal. The signals on I and Q are multiplied by a complex scrambling code. There are 2^{24} uplink scrambling codes. All uplink channels (except the PRACH) use short or long scrambling codes. The PRACH uses the long scrambling code. The modulation of both DPCCH and DPDCH is BPSK.

- **Spreading and Modulation of Downlink Physical Channels**. As in the case of uplink, the OVSF codes are used for channelization in the downlink. The I and Q branches have the same channelization or OVSF code. Spreading and modulation of DPCH, CPICH, Secondary CCPCH, PDSCH, PICH, and AICH are identical. Primary and Secondary SCHs are code multiplexed and transmitted simultaneously during the first 256 chips of each slot. The SCH is non-orthogonal to the other downlink physical channels. There are primary and secondary scrambling codes. Each cell is assigned one primary scrambling code. The primary CCPCH is transmitted using the primary scrambling code, while other downlink physical channels are transmitted using the primary or secondary scrambling code. The mixture of primary and secondary scrambling code is allowed for one CCTrCH. Different DPDCHs are multiplexed (in complex domain) and use the same scrambling code. The modulation scheme used is QPSK.
- **Power Control.** The basic power control operation is similar to the power control operation of IS-2000 shown in Figure 11.3.
  - In the uplink, the power control loop modifies the power of the DPCCH and DPDCHs with the same amount. The relative transmit power difference between DPCCH and DPDCH is conveyed to the UE by the network through higher layer signaling. The uplink inner-loop power control changes the UE transmit power to maintain the received uplink SIR at a given target SIR. The BSs in the Active Set estimate the SIR of the received uplink DPCH and generate TPC commands that are sent once per slot. The UE combines these commands into a single command using one of the two algorithms specified by the UTRAN. The UE adjusts the transmit power using the step size of 1 dB or 2 dB, which is controlled by the UTRAN.
  - In the downlink, the transmit power of DPCCH and the corresponding DPDCH are controlled simultaneously. The relative power difference between the DPCCH and DPDCH is not changed. The power control tries to keep the received downlink SIR at a given target SIR. An outer loop adjusts the target SIR independently for each connection. The UE estimates the received SIR and generates TPC commands. The network changes the transmit power in response to the received TPC commands from the UE in multiples of the minimum step size (1 dB mandatory, 0.5 dB optional).
- **Site Selection Diversity.** The Site Selection Diversity (SSD) is an optional macro-diversity method in soft handoff. The UE selects one cell from its Active Set as primary, while other cells are considered secondary. In the downlink, the power is transmitted only from the primary cell, reducing the interference caused by transmissions on all the

cells from the Active Set. SSD initiation and termination and assignment of temporary ID to each primary and secondary cell is carried out by upper layer signaling.

11.6.2 UTRA-TDD Features

Salient radio interface related features of UTRA-TDD are summarized next. Details of these features can be found in [120] and [127].

- **Time Slot Structure.** Each time slot has two data fields, one midamble field and one guard period (GP) field. The data fields have the information bits (from the transport channels) that have undergone multiplexing, interleaving, coding, and spreading. The midamble field carries training sequences. The GP compensates for timing inaccuracies, delay spread, and propagation delay, in case no timing advance mechanism is used, and helps power ramping. There are a total of 2560 chips in a time slot.
- **Dedicated Physical Channel.** There is one Dedicated Physical Channel (DPCH). Figure 11.9 outlines the physical channels of UTRA-TDD.

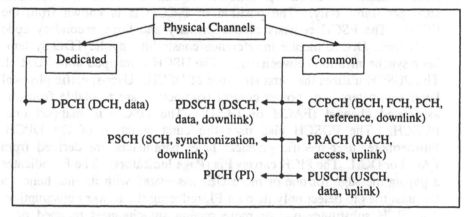

Figure 11.9: Physical Channels for UTRA-TDD.

The DPCHs carry data bits of the DCHs. For the DPCH, multi-code transmission with fixed spreading or a single code transmission with variable spreading can be considered. As with the FDD mode, the UTRA utilizes the TFCI for the TDD mode to support variable data rates and different services. The TFCI is not necessary for a simple service such as fixed rate speech. The TFCI is optionally transmitted in the UTRA TDD within the data fields of a burst, i.e., a time slot, adjacent to the midamble.

The vicinity of the TFCI to the midamble is important for reliable transmission of the TFCI. To implement the closed loop power control of the downlink DPCH, TPC commands are sent to the BS as the first two bits of the second data field.

- **Common Physical Channels.** Common Physical Channels are Common Control Physical Channels (CCPCH), Physical Random Access Channel (PRACH), Physical Synchronization Channel (PSCH), Physical Uplink Shared Channel (PUSCH), Physical Downlink Shared Channel (PDSCH), and Page Indicator Channel (PICH). The CCPCH contains BCH, PCH, and/or the FACH. The CCPCH is referred to as Primary CCPCH (P-CCPCH) when it is transmitted with reference power, has a known position in a frame, and carries BCH. The value of the fixed power of the P-CCPCH is broadcast and is used as a reference for other measurements by the terminal. The P-CCPCH uses a pre-defined channelization code and midamble and is transmitted in the first SCH time slot in a frame. The Secondary CCPCH (S-CCPCH) transmits FL common control information, e.g., FACH and PCH messages. The PRACH carries RACH. The terminals randomly transmit messages in the PRACH time slots to acquire access to the system. If two terminals transmit with the same code in the same time slot, a collision occurs. The TrCH called SCH is mapped onto PSCH. The PSCH utilizes one or two downlink slots per frame only. The position of P-CCPCH is known from the PSCH. The PSCH consists of one primary and three secondary code sequences. Due to mobile interference constraints, public TDD systems keep synchronization between cells. The USCH is mapped onto PUSCH. The PUSCH utilizes the burst structure of DPCH. User-specific physical layer parameters, e.g., power control parameters, are available from the associated channel (FACH or DCH). The DSCH is mapped onto PDSCH. The PDSCH also uses the burst structure of the DPCH. Furthermore, user-specific physical layer parameters are derived from FACH or DCH. The PICH carries PIs (Page Indicators). The PI indicates a paging message for one or more UEs associated with it, and, hence, a terminal tries to detect only its own PI, reducing the power consumption. The PICH substitutes one or more paging sub-channels mapped on a CCPCH.

- **Modulation and Spreading of Physical Channels.** After the processing of TrCHs is complete, physical channels are created. These physical channels undergo modulation and spreading. QPSK modulation is used. Interleaved and encoded data bits are converted into QPSK symbols. Each QPSK data symbol is spread using OVSF codes (a factor of 1, 2, 4, 8, or 16). The OVSF codes allow mixing of channels with different spreading factors in the same time slot while preserving orthogonality. A cell-specific scrambling code is multiplied with the proper length spread symbols.

- **Chip Rates.** Two chip rates available in the TDD mode are 3.84 Mcps (same as the FDD mode) and 1.28 Mcps.
- **Downlink Code Assignment.** The available CDMA codes can be assigned to either a single user or multiple users who are transmitting bursts in the same time slot. The maximum number of CDMA codes is sixteen. The actual number of CDMA codes, k, depends on individual SFs, interference, and service requirements.
- **Synchronization.** The primary code sequence is a generalized hierarchical Golay sequence, while the secondary synchronization codes are based on Hadamard sequences. The PSCH carries information such as the position of the time slot within the frame and location of the primary CCPCH.
- **Time Slot Utilization.** In a given radio frame, at least one time slot has to be allocated for the downlink and one for the uplink. Apart from this restriction, time slots can be distributed approximately equally or unequally between the uplink and the downlink to support symmetric or asymmetric data rate requirements.
- **Spreading for Downlink Physical Channels.** The spreading factor is limited to 16 used to reduce the complexity and cost of terminals. Multiple channelization codes can be used in parallel to support high data rates.
- **Spreading for Uplink Physical Channels.** The uplink transmission with one code and variable spreading factors (i.e., SF = 1, 2, 4, 8, or 16) or with multiple codes (a maximum of two different channelization codes) is allowed.
- **Training Sequences.** These sequences are used for channel estimation. The training sequences in the UTRA-TDD are obtained from a specific construction algorithm so that joint channel estimation for several users can be performed by a single cyclic correlator on the reverse link. On the downlink, the midambles for different users active in the same time slot are time-shifted versions of a single periodic basic code. Different cells use different periodic basic codes.
- **Handoff.** A TDD system does not use soft handoff. The terminal receives a list of neighboring cells and their important characteristics, e.g., frequency, codes, and midambles, from the BS so that appropriate measurements can be made. For a TDD target cell, the receive power of the beacon is measured. If the target cell is an FDD cell, E_c/I_o of the FDD CPICH is measured. In case of a GSM target cell, receive power of the GSM broadcast channel is measured.

- **Timing Advance Mechanism**. The UTRA-TDD employs a timing advance mechanism so that the bursts from all terminals are aligned with the overall frame structure. This mechanism compensates for distance related transmission delay for all the terminals. The TDD-CDMA can be operated with or without uplink synchronization. In both cases, a timing advance mechanism can be employed. This mechanism can also be disabled for an operation without the uplink synchronization.
- **Types of Bursts**. There are three types of data bursts based on how chips are distributed among data fields, midamble field, and the GP [127]. Burst Type 1 (BT 1) reserves 976 chips for each data field, 512 chips for the midamble field, and 96 chips for the GP. Burst Type 2 (BT 2) keeps 1104 chips for each data field, 256 chips for the midamble, and 96 chips for the GP. Burst Type 3 (BT 3) sets aside 976 chips for data field 1, 880 chips for data field 2, 512 chips for midamble, and 192 chips for the GP. BT1 allows fewer information chips but more midamble chips so that an improved channel estimate can be obtained for better FER performance. BT 2 provides higher information chip rate (and, hence, better spectrum efficiency). BT 3 has more chips for the GP, helping the PRACH to counteract timing inaccuracies in the initial access.
- **Transmit Diversity**. Downlink transmit diversity can be employed for the DPCH at the UTRAN. Examples of diversity schemes for DPCHs are Selective Transmit Diversity (STD) and Transmit Adaptive Antennas (TxAA). For the SCH, TSTD can be employed.
- **Power Control**. The power control for UTRA-TDD is similar to the power control operation for UTRA-FDD and IS-2000.
  - For the uplink, open loop power control is used. The transmit power for the DPCH is a function of factors such as target SIR, short-term path loss, long-term average path loss, quality of path loss measurement, interference, and a constant value set by higher layers of UTRAN. A higher layer outer loop adjusts target SIR.
  - For the downlink, the power control is based on SIR and is similar to the closed loop power control implemented for the FDD mode. The terminal compares estimated SIR of the received signal with a target SIR and sends a TPC command (up or down step) to the base station. The base station may change its transmit power for all of the terminal's physical channels based on the received TPC command. The power control step size may be varied by the BS based on the terminal's measured downlink interference. An outer loop controls the target SIR.

11.7 RADIO RESOURCE MANAGEMENT FOR UMTS

RRM for UMTS is discussed in several recent publications. Jorguseski [131] proposes a set of radio resource allocation algorithms that consists of resource estimation, QoS-aware scheduling, power, rate, and time-slot allocation, and call admission control. The discussion of the resource allocation considers access schemes such as UTRA-FDD and UTRA-TDD. Some general techniques for power, rate, and time slot allocation are suggested. The QoS in 3G air interfaces, radio access network, and core network is the focus of [132]. The UMTS QoS architecture is described by considering QoS parameters, traffic classes, the end-to-end data delivery model, and the mapping of end-to-end services to the services provided by the network elements of the UMTS. Koodli [133] also discusses QoS architecture and mechanisms for UMTS. Traffic classes defined in the UMTS are summarized, which include conversational class, streaming class, interactive class, and background class.

Since physical layers and methods of packet data transmission are different for the modes of the UMTS, RRM is also different for these modes. The RRM for TDD mode and the FDD mode is briefly discussed next. Only forward link RRM is considered here because reverse link is not expected to be the performance bottleneck.

RRM for the UTRA-FDD. The RRM for the UMTS-FDD is fundamentally similar to the RRM for IS-2000 in importance of forward link power, FDD approach, and use of Walsh codes (or channelization codes). Major resources to be managed are power and channelization codes. Understanding how packet data is transmitted often highlights important aspects of RRM.

In the downlink of the UTRA-FDD, data can be transmitted using one of two basic options, use of DCHs or use of a combination of a DCH and a DSCH [128]. In the first option, all users are allocated DCHs with variable spreading factors. Since there are only a limited number of spreading (or channelization) codes, the system may run out of spreading codes. To alleviate such a problem, a DCH can be released if there is no data activity for a certain time duration. A timer that controls this release requires proper setting. If the timer value is too long, the expensive radio resources are not used efficiently. A short timer value leads to frequent allocation and de-allocation of resources, increasing the signaling load and reducing the achievable capacity and throughput. It is also possible to downgrade the DCH by using a larger SF. In the second *option*, all users are allocated DCHs. Furthermore, some DSCHs are shared among active users. The DSCHs are allocated and de-allocated frequently. The DSCHs allow users to share resources on a need basis and alleviate the problem of spreading codes.

Two components of the RRM for UTRA-FDD are control of call admission and management of high data rate channels as shown in Figure 11.10.

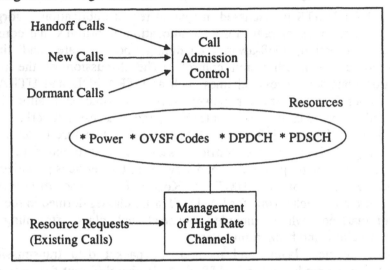

Figure 11.10: RRM for UTRA-FDD.

A call admission control mechanism accepts or rejects a new call request. When a new call arrives, it is admitted into the system if sufficient power and a channelization code are available. A channelization code corresponding to a low rate channel may be a good initial choice, because a low rate channel will require a small amount of power and will make more channelization codes available for future resource requests. Management of high data rate channels is required when an admitted call requires a high data rate channel to transfer a large amount of data. When such need arises, a lower spreading factor for a dedicated channel or a common shared channel can be used. Evaluation of the overall system performance for different configurations, i.e. DPCH or combined DPCH/PDSCH, can dictate a preferred option from the perspectives of capacity, throughput, and QoS. Note that an approximate idea of the power required for a high rate DPCH or PDSCH for a given user can be obtained by considering the power being used on the existing DCH and the differences between the existing DPCH and desired DPCH or PDSCH, e.g., data rates. Similar to the access of the high rate DPCH or PDSCH, the release of PDSCH or use of a low rate DPCH is also governed by factors such as amount of data, data arrival rate, and QoS. The time for which a PDSCH can be assigned to a given user (or a high rate DPCH is used) requires balancing of two competing requirements, the desire to offer a given user the best service and the need to share resources among users.

RRM for the UTRA-TDD. As mentioned before, a close look at the packet data transmission helps understand issues involved in the RRM. There are several approaches for packet data transmission in the UTRA-TDD mode [129]. Packet data transmission on common channels (e.g., FACH and RACH) has limited capacity, may experience collisions, and is inefficient because of the need for the UE identification in each TrBk. Packet data transmission on dedicated channels avoids collisions, is systematic, and has high capacity but can be somewhat inefficient because of the overhead associated with allocation and de-allocation of resources. An alternative to using common channels or dedicated channels is the concept of a shared channel operation. The RRC manages the use of shared channels. For all established, i.e., "admitted," packet calls, the transmission parameters are configured at the RNC, the Node B, and the UE. Actual physical resources are not allocated for all these packet calls. The RRC schedules the use of the shared resources for all users in a cell. The shared channel operation can potentially be done for both downlink and uplink. The allocation of the resources is time-limited, and hence, acknowledgement of the allocation message is avoided. The Radio Link Controller (RLC) takes care of any lost data, e.g., due to loss of an allocation message.

Two factors that distinguish the RRM of TDD from the RRM of FDD are management of time slots and management of frequency bands (Figure 11.11). In UTRA TDD, each time slot may be assigned to either uplink or downlink [127].

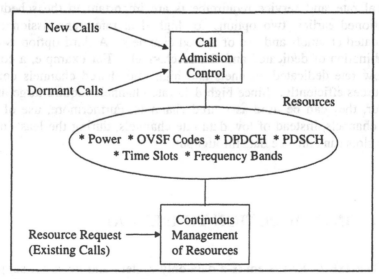

Figure 11.11: RRM for UTRA-TDD.

For a symmetric traffic, an approximately equal number of slots can be assigned to the downlink and the uplink. Widely different numbers of slots can be assigned to the downlink and the uplink for highly asymmetric traffic. Furthermore, the UTRA-TDD utilizes only one frequency band as opposed to the UTRA-FDD, which requires two frequency bands. The availability of two bands in the frequency spectrum for UTRA-TDD provides an opportunity for load or traffic balancing between the frequency bands using the technique of MCTA.

The RRM for the UTRA-TDD manages resources such as channelization codes (a maximum of sixteen for a given time slot), time slots, and power. A call admission algorithm allows a call to access the system if there is sufficient power and if a channelization code and a time slot are available. For all admitted calls, channelization codes and time slots need to be assigned according to service requirements and available resources. A DCA algorithm that considers interference measurements at the UE and the Node B can exploit the TDD advantages by proper management of time slots [130]. The network can create a priority list of time slots for each cell such that the lowest interference level is experienced when data is transmitted in proper time slots. In a CDMA system, low forward link interference requires low transmit power, potentially increasing the system capacity. A scheduling mechanism implemented at the network ensures that resources are efficiently shared among users to achieve high capacity and throughput. Since the DCA allows the network to create a priority list of time slots, scheduling of packets for different users is facilitated. Factors such as the amount of data, data arrival rate, and service requirements are important in the scheduling. As mentioned earlier, two options for high data rate transmission are use of dedicated channels and use of shared channels. A third option is the use of combination of dedicated and shared channels. For example, a combination of low rate dedicated channels and high rate shared channels could utilize resources efficiently. Since high data rate channels require large amounts of power, they can be used as shared channels. Furthermore, use of high data rate channels, instead of low data rate channels, during the least-interference time slots can enhance data throughput.

11.8   INTRODUCTION TO 1xEV-DO

The 1xEV-DO system is a data-only system and does not support voice. As a result, a separate carrier is required. In practice, a system based on IS-95, IS-2000, or UMTS could serve voice users, while the 1xEV-DO system would serve data users. The main goal of the 1xEV-DO system is to maximize throughput by discriminating users according to the radio channel conditions while maintaining desired quality of service. The standard refers to the mobile terminal or the data terminal as the Access Terminal (AT) and the base station as the Access Point (AP). Figure 11.12 illustrates logical system architecture of the 1xEV-DO. The Access Network Controller (ANC) controls multiple APs that are communicating with ATs. The Packet Control Function (PCF) communicates with multiple ANCs so that data can be transferred to the AT from one AP and ANC initially and another AP and ANC later in case the AP moves from the area served by one ANC into the area served by another ANC. In the physical system architecture, functions of the ANC and PCF may be distributed or centralized.

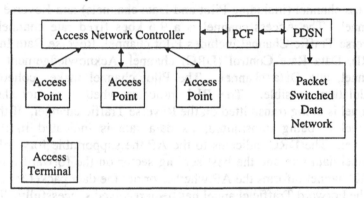

Figure 11.12: System Architecture for 1xEV-DO.

The common features between an IS-95 or IS-2000 system and a 1xEV-DO system are 1.2288 Mcps chip rate, link budgets, network plans, and RF designs for the AT and the AP. Salient Features of the 1xEV-DO system are summarized next. Figure 11.13 summarizes physical channels of IS-2000.

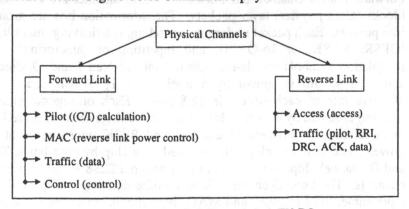

Figure 11.13: Physical Channels in 1xEV-DO.

- **Forward Link Channels.** The Forward Channel consists of Pilot channel, MAC channel, Traffic channel, and Control channel. The groups of 16 time slots are aligned to system time. During each slot, Pilot, MAC, and Traffic (or Control) channels are time-division multiplexed. All channels are transmitted at maximum sector power. Each slot is divided into two half slots, each with a pilot burst of duration 96 chips, centered at midpoint of the half slot. The MAC channel consists of two sub-channels: the Reverse Power Control (RPC) channel and the Reverse Activity (RA) channel. The RA channel transmits a Reverse Link Activity Bit (RAB) stream. The MAC channel bursts precede and follow each pilot burst and are of 64 chips.

- **Reverse Link Channels.** The reverse link channels consist of the Access channel (including Pilot and Data channels) and Reverse Traffic channel. The Access channel is a 9.6 kbps fixed rate channel. The Reverse Traffic Channel includes Pilot channel, Reverse Rate Indicator (RRI), Data Rate Control (DRC) channel, Acknowledgement (ACK) channel, and Data channel. The Pilot channel makes coherent de-modulation possible. The RRI indicates whether or not the Data channel is being transmitted on the Reverse Traffic channel. If the Data channel is being transmitted, its data rate is indicated in the RRI channel. The DRC indicates to the AP the supportable forward traffic channel data rate and the best serving sector on the forward link. The ACK channel informs the AP whether or not the data packet transmitted on the Forward Traffic channel has been received successfully. There is one acknowledgment channel bit in response to every Forward Traffic channel time slot. Otherwise, it is gated off. If the packet is not received correctly, a Negative Acknowledgment (NAK) is sent. The ACK bit is sent if the packet is received correctly.

- **Overall Processing in the Forward Link Physical Layer.** Both the Forward Traffic channel and Forward Control channel are encoded in blocks called physical layer packets. The information bits are grouped into packets. Each packet undergoes encoding, interleaving, modulation (QPSK, 8-PSK, or 16-QAM), and repetition or puncturing. The modulation symbols are de-multiplexed into sixteen I and Q streams, and each stream is spread by a distinct 16-ary Walsh code. The effective rate of each stream is 76.8 ksps. Each orthogonal channel (stream) is scaled by a gain of 1/4. The gain settings are normalized to a unity reference equivalent to unmodulated BPSK transmission at full power. Scaled Walsh chips are summed on a chip-by-chip basis. The I and Q channel chips are summed up to obtain 1.2288 Mcps on I and Q channels. The I and Q channel chips are time-division multiplexed with a preamble, Pilot channel, and MAC channel chips to create a sequence of chips for quadrature spreading using the Pilot PN sequence.

- **Overall Processing in the Reverse Link Physical Layer.** The reverse link physical layer is similar to the IS-2000 reverse link. In the reverse link, the modulation symbol rate is preserved at 307.2 ksps. The modulation is BPSK. The spreading bandwidth is 1.2288 MHz.
- **Rate Control and Power Control.** The data rate control on the forward link and fast power control on the reverse link help achieve target packet error rate (PER) in each link.
- **Soft Handoff.** There is no soft handoff on the forward link but rather fast cell switching. There is soft handoff on the reverse link. The AT receives data on the forward link from the strongest sector. Since the AP always transmits maximum sector power to the AT, it is possible to serve users near the cell edge. The soft handoff on the reverse link is useful because the AT power is relatively small, and diversity on the reverse link helps reduce the power requirements on the AT.
- **Timing.** There are stringent timing requirements in the system. Each sector's time base reference is synchronized to system time, within ±10 µsec of system time. Based on this time reference, time-sensitive transmission components such as PN sequence, time slots, and Walsh chips are derived.
- **User Distinction.** In the forward link, each user is assigned a MAC index. In the reverse link, the ATs use common short codes but long user-specific codes.
- **Data Rates.** Different data rates are obtained by using different numbers of physical layer packets and different modulation schemes on the forward link. The forward link data rate ranges from 38.4 kbps to 2.4576 Mbps. The reverse link utilizes different number of bits per packet to support a range of data rates. On the reverse link, the supported data rates are from 9.6 kbps to 153.6 kbps.
- **Layers Above Physical Layer.** The Radio Link Protocol (RLP) utilizes NAK. Sequencing of octets (instead of frames) removes complex segmentation and re-assembly issues if a retransmission frame cannot fit into the payload. The residual RLP FER is 10^{-6}.
- **System Architecture, Signaling, and Call Processing.** The system architecture is highly distributed. Call processing is simplified compared to IS-2000 systems. The call processing is characterized by combined traffic and paging channels, shortened RLP sync procedure, and absence of service negotiation. Thus, signaling delay is reduced, leading to faster call setup and improved data performance, e.g., reduced packet delay and larger throughput.

Forward Link Data Transmission. In the forward link, users share the system resources through time division multiplexing (TDM). Only one user is served at any given instant. The time slot is 1.66 ms long and consists of 2048 chips. During each time slot, Pilot, MAC, and Traffic (or Control) channels are time-division multiplexed. A forward link channel is always transmitted at full (i.e., maximum) power. Hence, it is possible to serve even a user near the cell edge at a high peak data rate. The number of bits transmitted to a user per slot depends on the data rate that the user requests and the network accepts. The AT measures pilot C/I every time slot for up to six strongest sectors in the Active Set of the reverse link soft handoff. The pilot C/I measurement is a reliable indication of the channel because all channels on the forward link are transmitted at full power. Since the pilots from all the sectors are transmitted at the same time at full power and since each pilot burst is preceded and followed by a MAC channel burst at full power, there is always some interference from each neighboring sector when pilot C/I is measured by the AT. This C/I measured by the AT can be considered to be the worst-case C/I for a traffic channel because interference from other sectors may or may not be present at the AT, depending upon the traffic loading and the time slots in use in the neighboring sectors, when the AT receives data on a traffic channel. Note that the pilot C/I is insensitive to system loading because no traffic is present when the pilot C/I is measured. In a typical IS-95 or IS-2000 system, pilot C/I measurement is sensitive to the traffic loading. The AT translates the C/I of the strongest sector into a data rate using an internal algorithm. The AT can choose a specific data rate option among several options available in the standard. These options use different modulation techniques and numbers of time slots to group the information bits.

The data rate option algorithm chooses the maximum rate at which data could be reliably received. The Hybrid-Automatic Repeat Request (H-ARQ) and RLP lead to much lower effective packet error rate. The AT conveys the requested data rate option to the AP on the DRC channel. The forward link transmission time slot is ½ slot before the reverse link transmission slot. After requesting the data rate option, the AT listens to the next time slot(s) (explained in detail later) for its data at the selected data rate option. Note that the AP must use the data rate option requested by the AT if it chooses to grant access to the AT. The AP could accept or reject the DRC channel requests based on its call admission and resource allocation algorithm. The AP transmits data to the AT according to the data rate option dictated by the DRC channel for a given user. When more than one slot is allocated to a user, four-slot interlacing is used for actual transmission to a user, leaving three interleaving time slots to other users. Such interlacing provides an opportunity to perform H-ARQ. In other words, if the AP receives a positive acknowledgment, i.e., if the intended physical layer packet has been correctly received, from the AT, the remaining un-transmitted time slots are de-allocated, i.e., released, and made available to other users. Note that the physical layer packet could contain the same information bits in an uncoded form and one or more coded forms. Thus, it is possible for the AT to retrieve distinct information bits of a physical layer packet before all the bits of the physical layer packet are fully received using all the allocated time slots. A time slot during which the AP does not send any data is called an *idle slot*. During the idle slot, no traffic or control channel data is transmitted; only the Pilot and MAC channels are transmitted.

Reverse Link Data Transmission. In the reverse link, the AT in the Connected state, i.e., when the AT has a traffic channel and is being fast power-controlled by the AP, can send data any time. Each AT is assigned a maximum allowable rate and a set of probabilities for calculation of the reverse link data rate. The rate of the data being sent by the AT to the AP is explicitly indicated on the RRI (Reverse Rate Indicator) channel. If the AT is in the Idle state, i.e., if it does not have a traffic channel, it first gains access to the system by requesting the AP through an access channel. Note that the call admission algorithm implemented at the AP accepts or rejects the AT request for an access to the system.

11.9 RADIO RESOURCE MANAGEMENT FOR THE 1xEV-DO SYSTEM

The 1xEV-DO system allocates time slots and MAC indices to all Active or Connected users. Since the channels are transmitted at full power, no forward link power management is required. In a typical CDMA system, e.g., an IS-2000 system, forward link power is the most valuable resource and must be carefully managed to achieve large capacity and throughput. However, in the case of a 1xEV-DO system, good management of resources such as time slots and MAC indices (and not resources such as the power) and data rate control is critical. Two major RRM components for the 1xEV-DO system are call admission and release control and time-slot management (Figure 11.14). These components are briefly discussed next.

Call Admission and Release Control. When a new call arrives (call origination), the system determines if it has a sufficient number of MAC indices to support the call and if it can provide a desired QoS. The system allows soft handoff on the reverse link, and one MAC index is required for power control of each link per sector in the active set. There are 64 MAC indices, with the indices 0 to 4 reserved and 5 to 63 available for traffic. Hence, there are 59 MAC indices that can be used to support traffic.

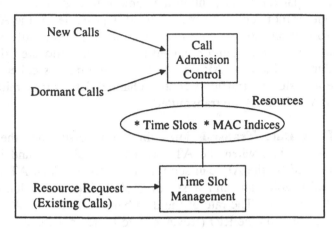

Figure 11.14: RRM in 1xEV-DO.

If the expected number of sectors per user (spu) is 2.0, the maximum number of distinct and active users that can be supported per sector is 59/2.0 = 29. If the spu is 1.5, a maximum of 39 active users per sector can be served. If the number of MAC indices is found to be the performance bottleneck, an attempt can be made to reduce handoff as much as possible while meeting the goal of call quality. If there are many Connected users in the system, the system may not be able to admit a new arriving call because of heavy utilization of time slots. Since a new arriving call cannot be queued or kept waiting for an arbitrarily long time, the system can block the call. In practice, the AP sends ConnectDeny message to the AT. Hence, two major criteria that control call admission in a 1xEV-DO system are lack of MAC indices and excessive call admission delay.

Call Release Control. Data packets often arrive in bursts, with a relatively long period of inactivity between the bursts. An AT does not need a MAC index during the inactivity periods. Hence, the AP can release the MAC index (previously reserved during the activity period) if there is no activity for certain time duration. Note that several ATs, already admitted into the system, may be experiencing data inactivity periods. If the system continues to reserve MAC indices for all the data calls it has admitted, the system may run out of MAC indices, preventing other ATs from making data calls. Such poor management of MAC indices may limit the system's ability to support high capacity and large throughput. A call release algorithm could implement a criterion that releases a MAC index if there is inactivity for specified time durations.

Time Slot Management. Data for all users from the PSDN eventually arrives at individual buffers at the serving AP. Hence, the AP has the knowledge of how much data needs to be transmitted for each user. The AP receives data rate option requests on the DRC channels from all the users at a predetermined frequency (i.e., every DRCLength time slots, where DRCLength is from 1 to 8). A packet scheduling algorithm determines which user should be served at a given time slot. *The packet scheduling algorithm is one of the key design components of a 1xEV-DO system, which effectively controls the overall system performance quantified by capacity, throughput, and user QoS.* According to the proportional fairness packet scheduler proposed in [134] and used in the performance evaluation in [135], the scheduler determines the next AT to be served based on data rate requests and the amount of data already transmitted to the ATs. The scheduler sends data to the AT that has the highest *DRC_rate/R*, where *DRC_rate* is the rate requested by the AT at each (slot assignment) decision time and *R* is the average rate received by the AT over a time window. Thus, a user with its requested rate closest to the peak is served. Scheduling is done at the time instant when a new transmission begins, while *R* is updated every slot.

When a user's data rate option with multi-slot transmission is accepted by the AP, appropriate time slots, i.e., time slots with a four-slot interlacing, are reserved for such a user. The reserved time slots are made available to other users only if a positive acknowledgment regarding the correct physical layer packet reception is received by the AP from the AT. As mentioned earlier, the forward link transmissions are ½ time slot ahead of the reverse link transmission. If DRCLength is 1, the AT sends data rate option preference every time slot and expects data arriving after ½ time slot to be according to the preference sent last time slot on the DRC channel. If DRCLength is 4, the data rate option preference is sent by the AT on the DRC channel during the four consecutive time slots. Beginning ½ time slot after sending the last DRC time slot, the AT expects data to start arriving at the just selected data rate option in the next four time slots. Once the AP starts transmitting according to the data rate preference, the data rate remains constant until all the allocated time slots are used. If the physical layer packet is not correctly received after all the time slots have been consumed, it is considered to be in error. If DRCLength is large, there are more opportunities for the AP to share the time slots among the data queues, e.g., prioritization of users or services and optimization of sector throughput or user throughput. When DRCLength is small, the AP has to decide fairly quickly how to allocate time slots. If the scheduling optimization is complex and needs many inputs or measurements, larger DRCLength is more appropriate. Smaller DRCLength is beneficial for better packet delay performance.

Simulation Methodology for RRM Evaluation of a 1xEV-DO System.
Performance evaluation of RRM of the 1xEV-DO system provides insight
into potential capacity, throughput, and packet delay for the system. The
RRM simulator framework discussed here is based on [135] and the RRM
simulator described in Chapter 3. Note that only the basic steps involved in
the simulation are mentioned below. A designer may need to simulate more
system features than the ones suggested here. The suggested simulation
mechanism can also be refined as more knowledge about the system itself is
gained, e.g., after field trials or commercial deployment. The simulator
discussed here does not model functions of upper layers, e.g., RLP and
TCP/IP protocols, and assumes that the upper layers will control the flow of
traffic such that data would arrive from the PDSN to the AP at a suitable rate.
A constant source data rate is used in the traffic generator of the simulator.
Major operations involved in the simulator are traffic generation, call
admission control, packet scheduling, call release control, and calculation of
performance metrics. Examples of useful performance metrics are average
number of traffic sessions, average number of Connected users (i.e., users
with MAC indices), connection blocking probability, distribution of user
perceived throughput (the ratio of the total number of bits transmitted and the
transmission time for a packet call), average sector throughput, E_c/I_o
distribution, data rate distribution, and packet delay distribution. All these
performance metrics can be found for different average numbers of traffic
sessions, average numbers of active users, and test cases that represent
different degrees of service penetration.

- Generate a new traffic session according to the exponential distribution
  with a mean of $D_{session}/N_{sessions}$ seconds. For example, if the average
  session duration, $D_{session}$, is 100 seconds and the average number of
  sessions, $N_{sessions}$, is 5, a new session is generated (on average) every 100/5
  = 20 seconds. Note that the exponential session inter-arrival time leads to
  Poisson arrival of traffic sessions. Classify the traffic session into a
  specific type of service, e.g., WWW or telnet, depending upon the test
  case.

- Execute the call admission algorithm for the traffic sessions of users that
  do not have any resources. In a simple form of the algorithm, the AP
  accepts the call request if it has sufficient number of MAC indices after
  reserving some MAC indices for reverse link handoff. If there are no
  MAC indices left, the AP could choose to queue the request for some time
  period. If no MAC index becomes available within such time period, the
  AP could ultimately deny the request for the system access. In the
  simulation, the ATs could retry accessing the system after some
  configurable time.

- For all accepted calls, make their states Connected and generate packets according to the data traffic models (as described in Chapter 3). After data packets are generated, they are sent to buffers.

- Assign MAC indices to the ATs in the Connected state. The ATs can be assigned E_c/I_o values according to the distribution generated using an off-line simulator. Another possibility is to locate the users uniformly in a sector (in a system layout with two tiers of out-of-cell interference) and model the effects of distance-based path loss, large-scale fading (log-normal shadow fading), and small-scale fading (Rayleigh or Ricean fading). Maximum sector power should be used in E_c/I_o calculation. Each user can be assumed to be in the same location for a specific time period.

- Determine data rate control options for Connected ATs based on a mapping between the E_c/I_o and data rate option. The frequency at which such determination is to be made is dictated by the choice of DRCLength. For the E_c/I_o and data rate option mapping, Table III of [136] or its variant modified for short-term fading can be used. See the standard for the available data rate options.

- Assign time-slots to the Connected ATs according to the packet scheduler. Multi-slot transmissions can be simulated in several ways. One way is to assume that the AT will need all the time slots to retrieve a packet accurately. Another way is to assign some probability of correct packet retrieval for a packet with a given data rate option and measured E_c/I_o.

- Perform a call release function if appropriate. In other words, release a MAC index for a traffic session if there is a long inactivity period. Note that the traffic session will trigger a new call request if there is data in the buffer but the AT does not have a MAC index, i.e., if it is not in Active state. Such a call request could be treated in the same way as the new call request or it could be prioritized using the temporary identifications assigned to all the calls.

- Observe performance metrics mentioned earlier.

- Repeat the simulation steps for a pre-determined time duration.

The simulator discussed above utilizes several modeling mechanisms, e.g., E_c/I_o measurement and multi-slot transmission, to reduce the simulation time. Such modeling may not yield very accurate results; however, if the simulation time is excessive, the simulator itself is not very useful. An efficient approach to evaluate system performance is to analyze major system functions, e.g., physical layer and multi-slot transmission, separately and mimic these functions in the RRM simulator using a suitable interface.

11.10 APPLICATION OF NEURAL NETWORKS AND FUZZY LOGIC TO RRM

Four handoff approaches are developed in this book using neural networks and fuzzy logic. These approaches include the generic fuzzy logic based approach, neural encoded fuzzy logic approach, unified candidacy based approach, and pattern classification based approach. These approaches can be generalized to two major components of the RRM, call admission and resource control for existing users. Figure 11.15 depicts the methodology for an RRM algorithm embodying a neural or fuzzy system.

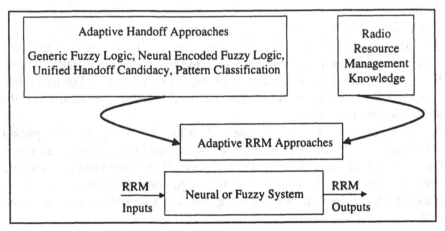

Figure 11.15: RRM using Neural Networks and Fuzzy Logic.

For call admission, potential inputs to the neural or fuzzy system are the type of session, type of service, measured queuing delay, available resources, and user priority, while one potential output of the neural or fuzzy system is the decision of the admission control mechanism. The type of the session may be new, handoff, or previously admitted but currently without radio resources. The handoff calls are usually prioritized over new calls by reserving some resources, e.g., power for a CDMA system, that are not accessible to new calls. The last type of session is a data session that has been inactive for a long time and does not have any dedicated radio resources. In IS-2000 terminology, such a session is considered to be in a dormant state. The service type may be a voice or a data service. A voice service cannot tolerate delay, while many data services can tolerate some delay. A data service may be queued until resources become available. An indicator of available resources is probably the most critical input. User priority could also be a useful input if users have subscribed to different levels of service.

The output of the neural or fuzzy system is admission or rejection of the request or queuing of the request.

For resource control, examples of inputs to the neural or fuzzy system are current resource usage, data related information, e.g., amount of data and arrival rate, service requirements, and user priority. Potential outputs are resource allocation or queuing decision. If there are no resources, the request is likely to be queued. When resources are available, the allocation parameters can be set according to the amount of data, data arrival rate, service requirements, and user priority. For example, the time duration for which a resource, e.g., a high rate channel, is reserved by a user is longer for a large amount of data, high data arrival rate, short delay requirements, and a user that has subscribed to a Gold-level service.

While creating rules for the neural or fuzzy system, desirable features such as large capacity and throughput and short packet delay should be considered. The methodology outlined for handoff in Chapter 4 can be used as a guide for development of an RRM algorithm.

11.11 SUMMARY

The three 3G and 3.5G cellular standards discussed in this chapter are IS-2000, UMTS, and 1xEV-DO. The IS-2000 and UMTS systems aim to provide increased voice capacity and enhanced data performance compared to 2G systems. The 1xEV-DO systems focus exclusively on data services. Radio interfaces and RRM of the standard are summarized. RRM for the 2G systems is relatively simple, because these systems are dominated by voice users. The variety of data services and co-existence of voice and data users make the resource management in the emerging systems complex. Efficient RRM can achieve the dual goals of high capacity and large throughput at satisfactory QoS levels. The framework created for handoff in this book can be exploited to develop high performance RRM algorithms for the emerging cellular systems.

REFERENCES

[1]   Theodore S. Rappaport, *Wireless Communications*. Prentice-Hall Inc, 1996.

[2]   George Liodakis and Peter Stavroulakis, "A Novel Approach in Handover Initiation for Microcellular Systems," Proc. 44th IEEE VTC, pp. 1820-1823, 1994.

[3]   Michel Mouly and Marie-Bernadette Pautet, *The GSM System for Mobile Communications*. Michel Mouly and Marie-Bernadette Pautet, 1992.

[4]   William C. Y. Lee, *Mobile Communications Design Fundamentals*. 2nd Ed. John Wiley & Sons Inc., 1993.

[5]   Gregory P. Pollini, "Trends in Handover Design," IEEE Communications Magazine, pp. 82-90, March 1996.

[6]   G. E. Corazza, D. Giancristofaro, and F. Santucci, "Characterization of Handover Initialization in Cellular Mobile Radio Networks," Proc. 44th IEEE VTC, pp. 1869-1872, 1994.

[7]   M. E. Anagnostou and G. C. Manos, "Handover Related Performance of Mobile Communication Networks," Proc. 44th IEEE VTC, pp. 111-114, 1994.

[8]   William C. Y. Lee, *Mobile Cellular Telecommunications*. 2nd Ed. McGraw Hill, 1995.

[9]   Vijay K. Garg and Joseph E. Wilkes, *Wireless and Personal Communications Systems*. Prentice-Hall Inc., 1996.

[10]   E. A. Frech and C. L. Mesquida, "Cellular Models and Handoff Criteria," Proc. 39th IEEE VTC, pp. 128-135, 1989.

[11]   W. R. Mende, "Evaluation of a Proposed Handover Algorithm for the GSM Cellular System," Proc. 40th IEEE VTC, pp. 264-269, 1990.

[12]   D. Munoz-Rodriguez and K. W. Cattermole, "Multicriteria for Handoff in Cellular Mobile Radio," IEE Proc., Vol. 134, pp. 85-88, 1987.

[13]   G. H. Senarath and David Everitt, "Comparison of Alternative Handoff Strategies for Micro-Cellular Mobile Communication Systems," Proc. 44th IEEE VTC, pp. 1465-1469, 1994.

[14]   Toshihito Kanai and Yukitsuna Furuya, "A Handoff Control Process for Microcellular Systems," Proc. 38th IEEE VTC, pp. 170-175, 1988.

[15]   Antonio J. and M. Ransom, "Handoff Considerations in Microcellular Systems Planning," Proc. PIMRC, pp. 804-808, 1995.

[16]   D. Munoz-Rodriguez, J. A. Moreno-Cadenas, M. C. Ruiz-Sanchez, and F. Gomez-Casaneda, "Neural Supported Handoff Methodology in Microcellular Systems, Proc. 42nd IEEE VTC, Vol. 1, pp. 431-434, 1992.

[17]   S. T. S. Chia, "The Control of Handover Initiation in Microcells," Proc. 41st IEEE VTC, pp. 531-36, 1991.

[18]   G. Falciasecca, M. Frullone, G. Riva, and A.M. Serra, "Comparison of Different Handover Strategies for High Capacity Cellular Mobile Radio Systems," Proc. 39th IEEE VTC, pp. 122-127, 1989.

[19]   Olle Grimlund and Bjorn Gudmundson, "Handoff Strategies in Microcellular Systems," Proc. 41st IEEE VTC, pp. 505-510, 1991.

[20]   A. N. Rosenberg, "Simulation of Power Control and Voice Channel Selection in Cellular Systems," Proc. 35th IEEE VTC, pp. 12-15, May 1985.

[21]   A. Salmasi and K. S. Gilhousen, "On the System Design Aspects of CDMA Applied to Digital Cellular and Personal Communication Network," Proc. 41st IEEE VTC, pp. 57-62, 1991.

[22]   J. Shapira, "Microcell Engineering in CDMA Cellular Networks," IEEE Trans. on Veh. Tech., Vol. 43, No. 4, pp. 817-825, Nov. 1994.

[23]   D. C. Cox and D. O. Reudink, "Layout and Control of High Capacity Systems," in *Microwave Mobile Communications*, Chapter 7: W.C. Jakes Jr. (ed.), Wiley, pp. 542-622, 1974.

[24]   R. Beck and F. W. Ho, "Evaluation and Performance of Field Strength Related Handover Strategies for Microcellular Systems," Third Nordic Seminar on Digital Land Mobile Radio Communication, pp. 12-15, September 1988.

[25]   R. Beck and H. Panzer, "Strategies for Handover and Dynamic Channel Allocation in Micro-Cellular Mobile Radio Systems," Proc. 39th IEEE VTC, pp. 178-185, 1989.

[26]   S. T. S. Chia and R. J. Warburton, "Handover Criteria for a City Microcellular Radio Systems," Proc 40th IEEE VTC, pp. 276-281, 1990.

[27]   M. Gudmundson, "Cell Planning in Manhattan Environments," Proc. 42nd IEEE VTC, pp. 435-438, April 1992.

[28]   M. Frullone, P. Grazioso, and G. Rive, "On the Optimum Allotment of Frequency Resources in Mixed Cellular Layouts," IEICE Trans. on Fundamentals of Electronic, Communications and Computer Sciences, Dec. 1992.

[29]   C. L. I, L. J. Greenstein, and R. D. Gitlin, "A Microcell/Macrocell Cellular Architecture for Low- and High-Mobility Wireless Users," IEEE JSAC, Vol. 11, No. 6, pp. 885-891, Aug. 1993.

[30]   H. Furukawa and Y. Akaiwa, "A Microcell Overlaid with Umbrella Cell System," Proc. 44th IEEE VTC, pp. 1455-59, 1994.

[31]   J. Worsham and J. Avery, "A Cellular Band Personal Communication Systems," Proc. 2nd Universal Personal Communications, pp. 254-257, 1993.

[32]   K. Ivanov and G. Spring, "Mobile Speed Sensitive Handover in a Mixed Cell Environment," Proc. 45th IEEE VTC, pp. 892-96, 1995.

[33]   James F. Whitehead, "Cellular Spectrum Efficiency via Reuse Planning," Proc. 35th IEEE VTC, pp. 16-20, 1985.

[34]   J. Naslund et al., "An Evolution of GSM," Proc. 44th IEEE VTC, pp. 348-52, 1994.

[35]   C. W. Sung and W. S. Wong, "User Speed Estimation and Dynamic Channel Allocation in Hierarchical Cellular System," Proc. 44th IEEE VTC, pp. 91-95, 1994.

[36]   L. B. Milstein et al., "On the Feasibility of a CDMA Overlay for Personal Communications Networks," IEEE JSAC, Vol. 10, No. 4, pp. 655-668, May 1992.

[37]   D. M. Grieco, "The Capacity Achievable with a Broadband CDMA Microcell Underlay to an Existing Cellular Macrosystem," IEEE JSAC, Vol. 12, No. 4, pp. 744-750, May 1994.

[38]   G. L. Lyberopoulos, J.G. Markoulidakis, and M.E. Anagnostou, "The Impact of Evolutionary Cell Architectures on Handover in Future Mobile Telecommunication Systems," Proc. 44th IEEE VTC, pp. 120-24, 1994.

[39]   N. W. Whinnett, "Handoff between Dissimilar Systems: General Approaches and Air Interface Issues for TDMA Systems," Proc. 45th IEEE VTC, pp. 953-57, 1995.

[40]   E. D. Re and P. Iannucci, "The GSM Procedures in an Integrated Cellular/Satellite System," IEEE JSAC, Vol. 13, No. 2, pp. 421-430, February 1995.

[41]   H. K. Lau, P. Liu, and K. C. Li, "Handoff Analysis for an Integrated Satellite and Terrestrial Mobile Switch over a Fading Channel," Proc. IEEE PIMRC, pp. 397-401, 1992.

[42]   L. R. Hu and S. S. Rappaport, "Personal Communications Systems Using Hierarchical Cellular Overlays," Proc. IEEE ICUPC, pp. 397-401, 1994.

[43]   K. G. Cornett and S. B. Wicker, "Bit Error Rate Estimation Techniques for Digital Land Mobile Radios," Proc. 41st IEEE VTC, pp. 543-548, 1991.

[44]   M. Hata and T. Nagatsu, "Mobile Location Using Signal Strength Measurements in Cellular Systems," IEEE Trans. on Veh. Tech., Vol. VT 29, No. 2, pp. 245-252, 1980.

[45]   G. D. Ott, "Vehicle Location in Cellular Mobile Radio Systems," IEEE Trans. Veh. Tech., Vol. VT-26, No. 1, pp. 43-36, Feb. 1977.

[46]   A. Gamst, R. Beck, R. Simon, and E. G. Zinn, "The Effect of Handoff Algorithms with Distance Measurement on the Performance of Cellular Radio Networks," Proc. International Conf. on Digital Land Mobile Radio Communications, pp. 367-373, 1987.

[47]   M. Greiner, L. Low, and R. W. Lorenz, "Cell Boundary Detection in the German Cellular Mobile Radio: System C," IEEE JSAC, Vol. SAC-5, No. 5, pp. 849-854, June 1987.

[48]   S. S. Rappaport, "Blocking, Handoff, and Traffic Performance for Cellular Communication Systems with Mixed Platforms," IEE Proceedings-I, Vol. 140, pp. 389-401, 1993.

[49]   Jack M. Holtzman and Ashwin Sampath, "Adaptive Averaging Methodology for Handoffs in Cellular Systems," IEEE Trans. on Veh. Tech., Vol. 44, No. 1, pp. 59-66, 1995.

[50]   A. Sampath and J. M. Holtzman, "Adaptive Handoffs Through Estimation of Fading Parameters," Proc. ICC, Vol. 2, pp. 1131-1135, May 1994.

[51]   R. Vijayan and J. M. Holtzman, "Sensitivity of Handoff Algorithms to Variations in the Propagation Environment," Proc. 2nd IEEE ICUPC, Vol. 1, pp. 158-162, Oct. 1993.

[52]   P. Dassanayake, "Effects of Measurement Sample on Performance of GSM Handover Algorithm," Electronic Letters, Vol. 29, pp. 1127-1128, June 1993.

[53]   P. Dassanayake, "Dynamic Adjustment of Propagation Dependent Parameters in Handover Algorithms," Proc. 44th IEEE VTC, pp. 73-76, 1994.

[54]   Ning Zhang and Jack Holtzman, "Analysis of Handoff Algorithms using both Absolute and Relative Measurements," Proc. 44th IEEE VTC, Vol. 1, pp. 82-86, 1994.

[55]   Gerhard Rolle, "The Mobile Telephone System C 450-a First Step Towards Digital," Proc. Second Nordic Seminar, Oct. 1986.

[56]   Chen-Nee Chuah and Roy D. Yates, "Evaluation of a Minimum Power Handoff Algorithm," Proc. IEEE PIMRC, pp. 814-818, 1995.

[57]   Chen-Nee Chuah, Roy D. Yates, and D. J. Goodman, "Integrated Dynamic Radio Resource Management," Proc. 45th IEEE VTC, pp. 584-88, 1995.

[58]   Mark D. Austin and Gordon L. Stuber, "Velocity Adaptive Handoff Algorithms for Microcellular Systems," IEEE Trans. Veh. Tech., Vol. 43, No. 3, pp. 549-561, Aug. 1994.

[59]   K. Kawabata, T. Nakamura, and E. Fukuda, "Estimating Velocity Using Diversity Reception," Proc. 44th IEEE VTC, pp. 371-74, 1994.

[60]   Mark D. Austin and Gordon L. Stuber, "Directed Biased Handoff Algorithm for Urban Microcells," Proc. 44th IEEE VTC, pp. 101-5, 1994.

[61]   Manjari Asawa and Wayne E. Stark, "A Framework for Optimal Scheduling of Handoffs in Wireless Networks," Proc. IEEE Globecom, pp. 1669-1673, 1994.

[62]   O.E. Kelly and V.V. Veeravalli, "A Locally Optimal Handoff Algorithm," Proc. IEEE PIMRC, pp. 809-813, 1995.

[63]   R. Rezaiifar, A.M. Makowski, and S. Kumar, "Optimal Control of Handoffs in Wireless Networks," Proc. 45th IEEE VTC, pp. 887-91, 1995.

[64]   H. Maturino-Lozoya, D. Munoz-Rodriguez, and H. Tawfik, "Pattern Recognition Techniques in Handoff and Service Area Determination," Proc. 44th IEEE VTC, Vol. 1, pp. 96-100, 1994.

[65]   V. Kapoor, G. Edwards, and R. Sankar, "Handoff Criteria for Personal Communication Networks," pp. 1297-1301, Proc. ICC, 1994.

[66]   M. Gudmundson, "Correlation Model for Shadow Fading in Mobile Radio Systems," Electronic Letters, Vol. 27, No. 23, pp. 2145-2146, Nov. 1991.

[67]   Yasuaki Kinoshita and Tomoharu Itoh, "Performance Analysis of a New Fuzzy Handoff Algorithm by an Indoor Propagation Simulator," Proc. 43rd IEEE VTC, pp. 241-245, 1993.

[68]   D. Munoz-Rodriguez, "Handoff Procedure for Fuzzy Defined Radio Cells," Proc. 37th IEEE VTC, pp. 38-44, 1987.

[69]   G.H. Senarath and David Everitt, "Performance of Handover Priority and Queuing Systems Under Different Handover Request Strategies for Microcellular Mobile Communication Systems," Proc. 45th IEEE VTC, pp. 897-90, 1995.

[70]   Sirin Tekinay and Bijan Jabbari, "Handover and Channel Assignment in Mobile Cellular Networks," IEEE Comm. Mag., pp. 42-46, Nov. 1991.

[71]   Sirin Tekinay and Bijan Jabbari, "A Measurement-Based Prioritization Scheme for Handovers in Mobile Cellular Networks," IEEE JSAC, Vol. 10(8), pp. 1343-1350, Oct. 1992.

[72]   P. O. Gassvik, M. Cornefjord, and V. Svensson, "Different Methods of Giving Priority to Handoff Traffic in a Mobile Telephone System with Directed Retry," Proc. 41st IEEE VTC, pp. 549-553, May 1991.

[73]   D. Giancristofaro, M. Ruggieri, and F. Santucci, "Queuing of Handover Requests in Microcellular Network Architectures," Proc. 44th IEEE VTC, pp. 1846-1849, 1994.

[74]   B. Eklundh, "Channel Utilization and Blocking Probability in a Cellular Mobile Telephone System with Directed Retry," IEEE Trans. Comm., Vol. COM-34, pp. 329-337, April 1986.

[75]   H. Panzer and R. Beck, "Adaptive Resource Allocation in Metropolitan Area Cellular Mobile Radio Systems," Proc. 40th IEEE VTC, pp. 638-645, May 1990.

[76]   Ming Zhang and Tak-Shing P. Yum, "Comparisons of Channel-Assignment Strategies in Cellular Mobile Telephone Systems," IEEE Trans. Veh. Tech., Vol. 38, No. 4, pp. 211-215, 1989.

[77]   E. D. Re, R. Fantacci, and L. Ronga, "A Dynamic Channel Allocation Technique Based on Hopfield Neural Networks," IEEE Trans. on Veh. Tech., Vol. 45, No. 1, pp. 26-32, February 1996.

[78]   M. Frodigh, "Optimum Dynamic Channel Allocation in Certain Street Microcellular Radio Systems," Proc. 42nd IEEE VTC, pp. 658-661, May 1992.

[79]   B. Narendran, P. Agrawal, and D. K. Anvekar, "Minimizing Cellular Handover Failures Without Channel Utilization Loss," Proc. IEEE Globecom, pp. 1679-1685, 1994.

[80]   D. J. Goodman, S. A. Grandhi, and R. Vijayan, "Distributed Dynamic Channel Assignment Schemes," Proc. 43rd IEEE VTC, pp. 532-535, May 1993.

[81]   R. D. Yates and C. Y. Huang, "Integrated Power Control and Base Station Assignment," IEEE Trans. on Veh. Tech., Vol. 44, No. 3, pp. 638-644, Aug. 1995.

[82]   S. V. Hanly, "An Algorithm for Combined Cell-Site Selection and Power Control to Maximize Cellular Spread Spectrum Capacity," IEEE JSAC, Vol. 13, No. 7, pp. 1332-1340, Sept. 1995.

[83]   Per-Erik Ostling, "High Performance Handoff Schemes for Modern Cellular Systems," Ph. D. Dissertation, Royal Institute of Technology, Sept. 1995.

[84]   Raymond C. V. Macario, *Cellular Radio-Principles and Design.* McGraw-Hill, 1993.

[85]   C. M. Simmonds and M. A. Beach, "Network Planning Aspects of DS-CDMA with Particular Emphasis on Soft Handoff," Proc. 43rd IEEE VTC, pp. 846-849, 1993.

[86]   R. C. Bernhardt, "Macroscopic Diversity in Frequency Reuse Radio Systems," IEEE JSAC, Vol. SAC-5, No. 5, pp. 862-870, 1987.

[87]   Szu-Wei Wang and Irving Wang, "Effects of Soft Handoff, Frequency Reuse, and Non-Ideal Antenna Sectorization on CDMA System Capacity," Proc. 43rd IEEE VTC, pp. 850-854, 1993.

[88]   Mikael Gudmundson, "Analysis of Handover Algorithms," Proc. 41st IEEE VTC, pp. 537-54, 1991.

[89]   Li-Xin Wang, *Adaptive Fuzzy Systems and Control*. PTR Prentice Hall, 1994.

[90]   E. H. Mamdani, "Applications of Fuzzy Algorithms for Simple Dynamic Plant," Proc. IEE, Vol. 121, No. 12, pp. 1585-1588, 1974.

[91]   Simon Haykin, *Neural Networks: A Comprehensive Foundation*. Prentice-Hall, 1994.

[92]   D. Nguyen and B. Widrow, "Improving the Learning Speed of 2-layer Neural Networks by Choosing Initial Values of the Adaptive Weights," International Joint Conference of Neural Networks, Vol. 3, pp. 21-26, July 1990.

[93]   R. Vijayan and J.M. Holtzman, "Analysis of Handover Algorithm Using Nonstationary Signal Strength Measurements," Proc. IEEE GLOBECOM, Vol. 3, pp. 1405-1409, Dec. 1992.

[94]   M. D. Austin and G. Stuber, "Cochannel Interference Modeling for Signal Strength Based Handoff Analysis," Electronic Letters, Vol. 30, pp. 1914-1915, Nov. 1994.

[95]   Andrew J. Viterbi, Audrey M. Viterbi, Klein S. Gilhousen, and Ephraim Zehavi, "Soft Handoff Extends CDMA Cell Coverage and Increases Reverse Link Capacity," IEEE JSAC, Vol. 12, No. 8, pp. 1281-1288, 1994.

[96]   D. Hong and S. S. Rappaport, "Traffic Model and Performance Analysis for Cellular Mobile Radio Telephone Systems with Prioritized and Nonprioritized Handoff Procedures," IEEE Trans. Veh. Tech, Vol. VT-35, No. 3, pp. 77-92, Aug. 1986.

[97]   Hai Xie and Simon Kuek, "Priority Handoff Analysis," Proc. 43rd IEEE VTC, pp. 855-858, 1993.

[98]   Y-B. Lin, L-F. Chang, and A. Noerpel, "Modeling Hierarchical Microcell/Macrocell PCS Architecture," Proc. ICC, Vol. 405-9, 1995.

[99]   S. A. El-Dolil, W.C. Wong, and R. Steele, "Teletraffic Performance of Highway Microcells with Overlay Macrocell," IEEE JSAC, Vol. SAC-7, pp. 71-78, 1989.

[100]   X. Lagrange and P. Godlewski, "Teletraffic Analysis of Hierarchical Cellular Networks," Proc. 45th IEEE VTC, pp. 882-86, 1995.

[101]   P. Harley, "Short Distance Attenuation Measurements at 900 MHz and 1.8 GHz Using Low Antenna Heights for Microcells," IEEE JSAC, Vol. 7, pp. 5-11, Jan. 1989.

[102]   J-E. Berg, R. Bownds, and F. Lotse, "Path Loss and Fading Models for Microcells at 900 MHz," Proc. 42nd IEEE VTC, pp. 666-671, May 1992.

[103]   C. Loo and N. Secord, "Computer Models for Fading Channels with Applications to Digital Transmission," IEEE Trans. Veh. Tech., Vol. VT-40, pp. 700-707, 1991.

[104]   G. N. Senarath and David Everitt, "Combined Analysis of Transmission and Traffic Characteristics in Micro-Cellular Mobile Communication Systems," Proc. 43rd IEEE VTC, pp. 577-580, 1993.

[105]   G. N. Senarath and David Everitt, "Combined Analysis of Transmission and Traffic Characteristics in Micro-Cellular Mobile Communication Systems, Emphasis: Handoff Modeling," Proc. Seventh Australian Teletraffic Conf., Nov. 1992.

[106]   D. Everitt and D. Manfield, "Performance Analysis of Cellular Mobile Communication Systems with Dynamic Channel Assignment," IEEE JSAC, Vol. 7(8), pp. 22-34, Oct. 1989.

[107]   Ning Zhang and Jack Holtzman, "Analysis of a CDMA Soft Handoff Algorithm," Proc. IEEE PIMRC, pp. 819-23, 1995.

[108]   Toshihito Kanai, Masanori Taketsugu, and Seiji Kondo, "Experimental Digital Cellular System for Microcellular Handoff," Proc. 38th IEEE VTC, pp. 170-175, 1988.

[109]   TIA TR45.5 Working Group 4, "Proposed Text for Annex 2," March 1998 Contribution, Steven D. Gray (Chair), 1998.

[110]   TR 101 112 V3.2.0 (1998-04), "Selection Procedures for the Choice of Radio Transmission Technologies of the UMTS," 1998.

[111]   J. Ho, Y. Zhu, and S. Madhavapeddy, "Throughput and Buffer Analysis for GSM General Packet Radio Service (GPRS)," Proc. of IEEE Wireless Communications and Networking Conference (WCNC), Vol. 3, pp. 1427-1431, Sept. 1999.

[112]   Shuang Deng, "Empirical Model of WWW Document Arrivals at Access Link," Proc. ICC, Vol. 3, pp. 1797-1802, 1996.

[113]   Nishith D. Tripathi, "Data Traffic Models for Simulation Based Analysis of Wireless Systems," Approved for External Publication, Nortel Report, 2000.

[114]   Thomas Kunz, Thomas Barry, Xinan Zhou, James P. Black, and Hugh M. Mahoney, "WAP Traffic: Description and Comparison to WWW Traffic," Proceedings of the Third ACM International Workshop on Modeling, Analysis, and Simulation of Wireless and Mobile Systems (MSWiM 2000), Boston, USA, August 2000.

[115]   IS-2000 Standard.

[116]   David W. Paranchych, "On the Performance of Fast Forward Link Power Control in IS-2000 CDMA Networks," Proc. IEEE WCNC, Chicago, USA, Sept. 23-28, 2000.

[117]   Nishith D. Tripathi, Jeffrey H. Reed, and Hugh F. VanLandingham, "A New Class of Fuzzy Logic Based Adaptive Handoff Algorithms for Enhanced Cellular System Performance," Proc. Wireless '97, Vol. 1, pp. 145-164, July 1997.

[118]   A. Murase, I. C. Symington, and E. Green, "Handover Criterion for Macro and Microcellular Systems," Proc. 41st IEEE VTC, pp. 524-530, 1991.

[119]   Ashvin Chheda, "A Performance Comparison of the DS-CDMA IS-95B and IS-95A Soft Handoff Algorithms," Proc. 49th IEEE VTC, Vol. 2, pp. 1407-1412, 1999.

[120]   3GPP Technical Specification Group (TSG), TS 25, v3.0.0, 1999.

[121]   1xEV-DO Standard, 2000.

[122]   K. W. Richardson, "UMTS Overview," Electronics and Communication Engineering Journal, June 2000, pp. 93-100.

[123]   Yiping Wang, David W. Paranchych, and Ashvin H. Chheda, "Power Control Methods in Dedicated Control Channel with Discontinuous Transmission in IS-2000 Systems," Proc. IEEE PIMRC, Sept. 2000.

[124]. A. Hiroike, F. Adachi, and N. Nakajima, "Combined Effects of Phase Sweeping Transmitter Diversity and Channel Coding," IEEE Trans. Veh. Tech., Vol. 41, No. 2, pp. 170-176, May 1992.

[125]   Sarvesh Sharma and Ahmad Jalali, "Traffic Allocation and Dynamic Load Balancing in a Multiple Carrier Cellular Wireless Communication System," United States Patent No.: 6,069,871, Nortel Networks, May 30, 2000 (Filed: March 6, 1998).

[126]   Josef F. Huber, Dirk Weiler, and Hermann Brand, "UMTS, the Mobile Multimedia Vision for IMT-2000: A Focus on Standardization," IEEE Communications Magazine, Sept. 2000, pp. 129-136.

[127]   Martin Haardt, Anja Klein, Reinhard Kohn, Stefan Oestreich, Marcus Purat, Volker Sommer, and Thomas Ulrich, "The TD-CDMA Based UTRA TDD Mode," IEEE JSAC, Vol. 18, No. 8, August 2000, pp. 1375-1385.

[128]   Paolo Goria and Davide Sorbara, "System Performance Evaluation of Packet Data Transfer in UTRA-FDD Standard," Proc. IEEE RAWCON, pp. 95-98, 2000.

[129]   Christina GeBner, Reinhard Kohn, Jorg Schniedenharn, and Armin Sitte, "UTRA TDD Protocol Operation," Proc. 11th IEEE PIMRC, Vol. 2, pp. 1226-1230, 2000.

[130] Yiannis Argyropoulos, Scott Jordan, and Srikanta P. R. Kumar, "Dynamic Channel Allocation in Interference-Limited Cellular Systems with Uneven Traffic Distribution," IEEE Trans. on Veh. Tech., Vol. 48, No. 1, January 1999, pp. 224-232.

[131] L. Jorguseski, E. Fledderus, J. Farserotu, and R. Prasad, "Radio Resource Allocation in Third-Generation Mobile Communication Systems," IEEE Pers. Commun. Mag., pp. 117-123, Feb. 2001.

[132] S. Dixit, Y. Guo, and Z. Antoniou, "Resource Management and Quality of Service in Third-Generation Wireless Networks," IEEE Pers. Commun. Mag., pp. 125-133, Feb. 2001.

[133] R. Koodli and M. Puuskari, "Supporting Packet-Data QoS in Next Generation Cellular Networks," IEEE Pers. Commun. Mag., pp. 180-188, Feb. 2001.

[134] F. Kelly, "Charging and Rate Control for Elastic Traffic," European Transactions on Telecommunications, Vol. 8, pp. 33-37, 1997.

[135] A. Jalali, R. Padovani, and R. Pankaj, "Data Throughput of CDMA-HDR a High Efficiency-High data Rate Personal Communication Wireless System," Proc. 51st IEEE VTC, Vol. 3, pp. 1854-58, Spring 2000.

[136] Paul Bender, Peter Black, Matthew Grob, Roberto Padovani, Nagabhushana Sindhushyana, and Andrew Viterbi, "CDMA/HDR: A Bandwidth Efficient High Speed Wireless Data Service for Nomadic Users," IEEE Communications Magazine, Vol. 38, No. 7, pp. 70-77, July 2000.

[137] Nishith D. Tripathi, Jeffrey H. Reed, and Hugh F. VanLandingham, "Fuzzy Logic Based Adaptive Handoff Algorithms for Microcellular Systems," Proc. 49th IEEE VTC, Vol. 2, pp. 1419-1424, 1999.

[138] Nishith D. Tripathi, Jeffrey H. Reed, and Hugh F. VanLandingham, "Adaptive Handoff Algorithms for Cellular Overlay Systems," Proc. 49th IEEE VTC, Vol. 2, pp. 1413-1418, 1999.

[139] Nishith D. Tripathi, Jeffrey H. Reed, and Hugh F. VanLandingham, "Handoff in Cellular Systems," IEEE Pers. Commun. Mag., Vol. 5, No. 6, pp. 26-37, Dec. 1998.

[140] Nishith D. Tripathi, Jeffrey H. Reed, and Hugh F. VanLandingham, "Pattern Classification Based Handoff Using Fuzzy Logic and Neural Nets," Proc. IEEE ICC, Vol. 3, pp. 1733-1737, 1998.

[141] Nishith D. Tripathi, Jeffrey H. Reed, and Hugh F. VanLandingham, "An Adaptive Direction Biased Handoff Algorithm With Unified Handoff Candidate Selection Criterion," Proc. 48th IEEE VTC, Vol. 1, pp. 127-131, 1998.

INDEX